251
Amazing Facts of Mathematics

251
Amazing Facts of Mathematics

Rajesh Kumar Thakur

DIVE IN THE OCEAN OF KNOWLEDGE
Ocean Books

Ocean Books Pvt. Ltd.
ISO 9001:2015 Publishers

Published by
Ocean Books (P) Ltd.
4/19 Asaf Ali Road,
New Delhi-110 002 (INDIA)
e-mail: info@oceanbooks.in

ISBN 978-81-8430-134-2
251 Amazing Facts of Mathematics
by Rajesh Kumar Thakur

Edition
2024

Price
₹ 400.00 (Rupees Four Hundred only)

Printed at
Narula Printers, Delhi

Dedicated to

My Grandfather
Late Shri VishnuDeo Thakur
&
My Grandmother
Mrs. Sumitra Thakur

Foreword

Mathematics is not all about calculation. Its inner beauty can be judged once you dive the deep ocean of mathematics and taking out some interesting pearls for self satisfaction. The deeper you dive, the chance of getting pearls of hig quality will emerge better. Newton had done a great service to mankind by his discovery but he was so humble that he once said:–*I don't know what I may appear to the world; but to myself I seem to have only like a boy playing on the seashore and diverting a smoother peobble or a prettier shell than ordinary, whilst the great ocean of truth lay all undiscovered before me.*

Pythangoras sacrified 100 oxen, when he discovered the famous Pythagoras Theorem: Archimedes ran out naked when he found the famous Archimedes Principle and the great mathematician Isaac Newton sacrificed his love to go deeper in mathematical research and never married. This is not all, the all times great mathematical genius Leonhard Euler lost the sight of his right eye at the age of 28 and became completely blind around 60 but his love with mathematics never took a back seat and he published over 400 papers. The history of mathematics is very informative and interesting. In order to understand mathematics better, you must have the knowledge of its history.

The present book written by Mr. Rajesh Thakur is a collection of 251 such informative mathematical pearls which might have been collected upon several years of extensive research. This book touches not only the history of ancient mathematics but it also focused the research done in mathematics till 2011. The information collected here are so interesting and informative that lover of mathematics and researcher will find this books of immense help.

I congratulate Mr. Thankur for writing such informative book which brings the different branch of mathematics under a single roof. I wish this book will make a record in popularity and budding mathematician and lover of mathematics will be able to see other sight of mathematical beauty which is rare to find in any book. My best wishes are with the publisher who took pain in bringing such informative book and also to Mr. Thakur who did a better service by making the collection public.

Dr. JJ Rawal
President
The Indian Planetary Society

Preface

"The essence of mathematics is not to make simple things complicated, but to make complicated things simple"

—S. Gudder

Mathematics is an idea and it is to be created by the individual. Mathematics is not a subject to keep a distance but you should find the way to minimize the distance because you cannot keep mathematics away from your life. Even the nature exhibits the love of Mathematics. Fibonacci number, Equiangular Spiral, Golden rectangle can be found in nature.

It is sad that mathematics has got the label of subject meant for persons having higher IQ but if you go through the history of mathematics you will find that the scientist of great repute like Newton, Einsteen were not brilliant during childhood. C. F. Gauss; a noted mathematician called mathematics the queen of all subject and this looks true when you see the people from different background later embracing mathematics. Fermat (a lawyer), Galileo (a doctor), Elan Turing (Runner), Mersenne (Priest), Napier (Businessman), Christopher Wren (Architect), Charles Dodgson (Novelist) in later phase of life became a lover of mathematics.

This book has the collection of 251 very interesting tid-bits from different fields under one roof making this equally knowledgeable, and entertaining.

It took me more than 7 years to collect the rarest tid-bits for my avid readers. I have tried to incorporate story of some mathematical symbols and a few words used in mathematics and their origin.

I do hope the long demand of the readers to have such books on mathematics will be met.

I would like to extend my gratitude and acknowledge the support of EF Robertson of Mac Tutor Archives and Jeff Millers, a teacher in Florida. I have taken the help of hundreds of websites in order to prepare the contents and in due course I have taken the permission to use some of the contents from their web pages. I have mentioned the name of few in Bibliography itself.

I shall not forget the support and valuable suggestions of my young friends Mr. Gaurav, Ms. Meena, Ms. Meenakshi, Mr. Satyender and Mr. Satyanarayan Arya.

I have tried to put this book up-to-date and many of the events mentioned in this book has happend in the year 2011 itself.

Any suggestion for improvement of the book will be highly appreciated.

—Rajesh Kumar Thakur
rkthakur1974@gmail.com

The Ten Commandments of Mathematics

1. You shall read the problem...carefully.
2. Whatsoever you do to one side of the equation, do that also to the other.
3. You must use your common sense, else you will have a flagpole 9,000 feet high and even fathers younger than sons.
4. You shall ignore the teachings of false prophets to do all the work in your head.
5. When you do not know, don't look it up. If the problem still haunts you ask, the knowledgeable teacher around you.
6. You shall master each step before putting your heavy foot down on the next.
7. Your correct answer does not prove that you have worked that problem correctly.
8. You shall first see that you have copied your problem correctly, before bearing false witness that the answer book is incorrect.
9. You shall look back even into your youth and remember your arithmetic.
10. You shall learn, read, write, speak, and listen correctly in the language of mathematics, and verily A's and B's shall follow that even up to graduation.

Contents

1

Numbers

1. There are only 2 numbers 1 and 81 whose square root is equal to the sum of its digits

$$\sqrt{1} = 1, \sqrt{81} = 9 = 8 + 1$$

2. Earlier 1 was not considered as a number. Till the 15th century most of the mathematician Pacioli (1494), Kibel (1514), Baker (1568) excluded unity as number from their book. But thanks to Stevin, who dared to consider 1 as a number.

3. 7 is a Holy number. There are 7 days in creation, 7 days in the week, 7 phases of the moon, every 7th year was sabbatical, and 7 times 7 years was the jubilee. 7 is the smallest number of integer-sided rectangles that tile a rectangle so that no 2 rectangles share a common length. 7 is also a Mersenne Prime.

$7 = 2^3 - 1$.

So many features associated with 7 make it a very special number. The Christians consider 7 as a mystical number.

4. The Babylonians had an advanced number system, in some ways more advanced than our present systems. It was a positional system with a base of 60 rather than the system with base 10 in widespread use today. The Babylonians divided the day into 24 hours, each hour into 60 minutes, each minute into 60 seconds. The Babylonians used the formula:—

$$ab = [(a + b)^2 - a^2 - b^2]/2$$

to make multiplication easier. Even better is their formula

$$ab = [(a + b)^2 - (a - b)^2]/4$$

The Babylonians did not have an algorithm for long division. Instead they based their method on the fact that $a/b = a \times (1/b)$

This shows that Babylonians had the knowledge of squaring a number and also the idea of multiplicative inverse.

Multiplicative inverse of 'a' is $\frac{1}{a}$

5. **Cube Numbers:-** The numbers which can be represented by three dimensional cubes are called cubic number. 1, 8, 27, 64, 125...are cubic numbers which are obviously the cubes of 1, 2, 3, 4, 5,....

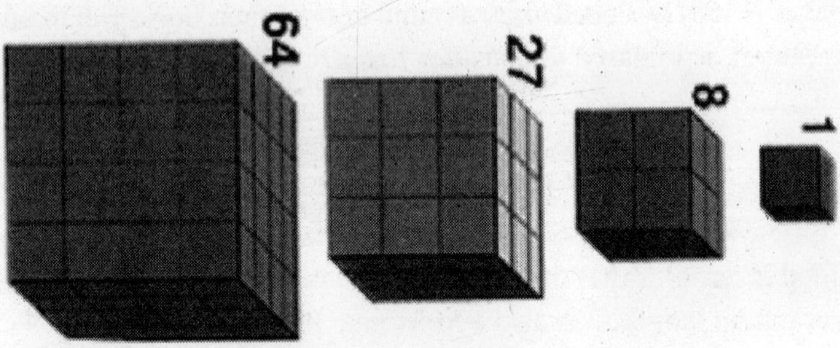

Properties of Cubic Numbers:-

- 1^3=1 = first odd number
  2^3=8=3+5 = sum of next two odd numbers
  3^3=27=7+9+11= sum of next three odd numbers
  4^3=64=13+15+17+19 = sum of next four odd numbers
  5^3=125=21+23+25+27+29 = sum of next five odd numbers
- Between 1 and 100 there are only two numbers 1 and 64 that are also square numbers.
- If C_1, C_2, C_3 ....are the first, second, third... cubic number; then they exhibit a unique property:-
  $C_1 = (T_1)^2$

$C_1 + C_2 = 1 + 8 = (T_2)^2$
$C_1 + C_2 + C_3 = 1 + 8 + 27 = 36 = (T_3)^2$
$C_1 + C_2 + C_3 + C_4 = 1 + 8 + 27 + 64 = 100 = (T_4)^2$
Where $T_1, T_2, T_3, \ldots$ are the triangular numbers.

6. Every integer greater than 20161 can be written as the sum of two Abundant numbers. A number **n** is called *Abundant* if the sum of its divisors is more than **2n.**

Example : 12 is the first Abundant number. Factor of 12 is 1, 2, 3, 4, 6 and 12 and its sum is 1+2+3+4+6+12 =28. The smallest odd Abundant number is 945.

7. We all know π =22/7 =3.142857142857…

Take the number repeated after decimal i.e. 142857 and multiply it with different numbers from 1 to 6. You shall always find the number repeated in a circular order.

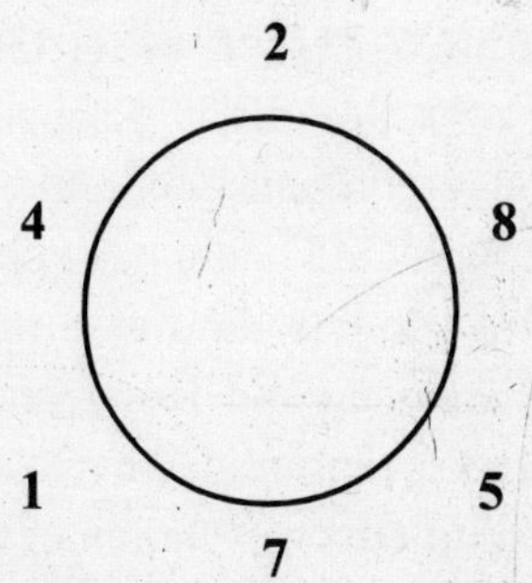

142857×1=142857
142857×2=285714
142857×3=428571
142857×4=571428
142857×5=714285
142857×6=857142

When the above result is written in tabular forms, the column wise or row-wise sum is always 27

Number Digit	1	4	2	8	5	7	Total
1	1	4	2	8	5	7	27
2	2	8	5	7	1	4	27
3	4	2	8	5	7	1	27
4	5	7	1	4	2	8	27
5	7	1	4	2	8	5	27
6	8	5	7	1	4	2	27
Total	27	27	27	27	27	27	

If we square the last three digits of 142857 and subtract the square of first three digits we get back the cyclic permutation of the number $857^2 - 142^2 = 714285$.

8. So far mathematicians have discovered only 24 perfect numbers. The first 6 are 6, 28, 496, 8128, 33550336, 8589869056 and the last one is of 12003 digits. Earlier, such numbers had mystical and religious significance for philosophers. Saint Augustine writes that although God could have created the world all at once, he preferred to take six days as 6 is symbolised by the perfect numbers whereas the old Testament argued that the perfection of the Universe is represented by 28, the number of days it takes the moon to complete one rotation of the earth.

Perfect number is a number whose sum of factors excluding itself is equal to the number itself.

6 has its factor 1, 2, 3 and 1+2+3 =6

28 has its factor 1, 2, 4, 7, 14 and 1+2+4+7+14=28

9. The early Chinese mathematicians used black colour to show the negative number and red colour to show positive number.

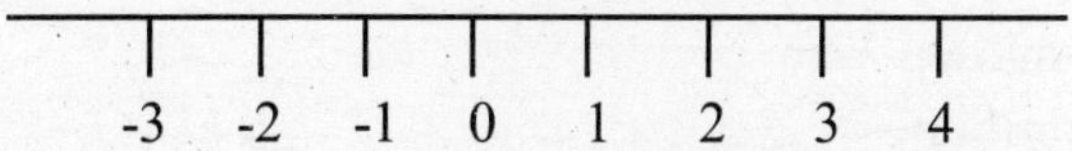

10. The first reference of Hindu-Arabic numerals was found in an Arabic treatise of Al-Khwarizmi. The original Arabic text is lost but its Latin translation *Algoritmi de numero Indorum* helped to introduce a new word—*algorithm* in mathematics.

Al-Khwarizmi

11. The prime number 73, 939, 133 has a very strange property. If you keep removing a digit from the right hand end of the number, each of the remaining numbers is also prime. It is the largest number known with this property. Take a look:

73, 939, 133 is a prime number
73, 93913 is a prime number
73, 939 1 is a prime number
73, 939 is a prime number
7, 393 is a prime number
739 is a prime number
73 is a prime number
7 is a prime number

12. We know there are many big—big numbers such as

10^6=million

10^9=billion

10^{12}=trillion

10^{30}=nonillion

10^{60}=novemdecillion,

10^{63}=vegintillion

10^{100}=googol

10^{googol}=googolplex

But can you believe 1 googolplex number will fill 1.25×10^{94} volumes if we assume that a normal 300 pages book contains approximately 800000 digits.

13. If the cube of an integer n is required, then let the first number a=n, d=2n, and t_n=n, and sum it up to n terms.

For example—

$7^3 = 7 + 21 + 35 + 49 + 63 + 77 + 91 = 343$

Here, a =7, n =7, d =14

This formula was given by Mahavira, an Indian mathematician, and it is based on Arithmetic Progression.

14. **Hexagonal Number:** Those numbers which form a shape of hexagon are called hexagonal numbers. 1, 6, 15, 28, 45.... are the few examples of hexagonal numbers.

Hexagonal numbers are of the form n (2n-1).

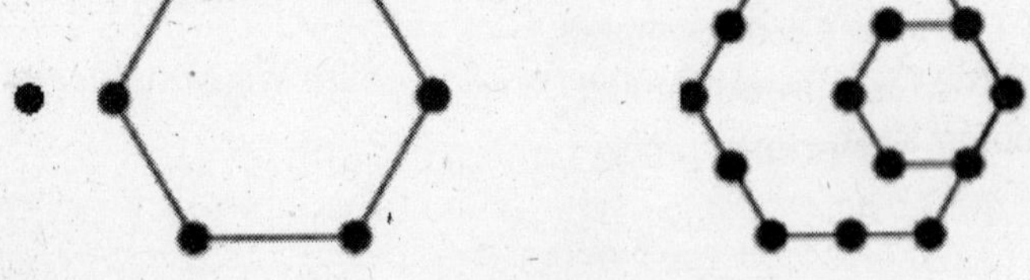

1 6 15

Properties:

- Every hexagonal numbers is a triangular number.
- 1, 7, 19, 37, 61, 91... are the centered hexagonal numbers.

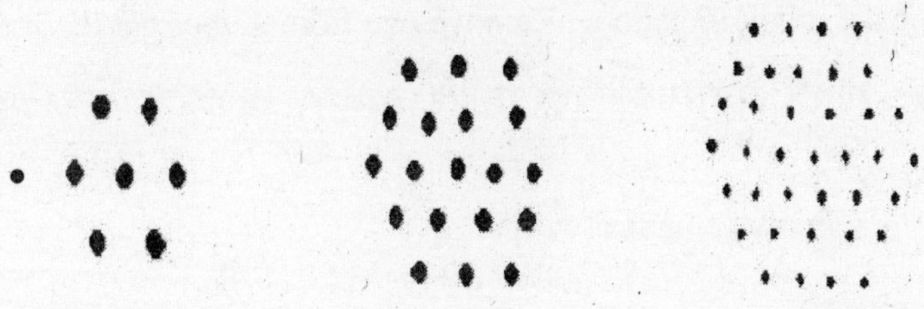

1 7 19 37

- 11 and 26 are the only numbers that can be represented by the sum using the maximum possible of six hexagonal numbers.

$11 = 1 + 1 + 1 + 1 + 1 + 6$
$26 = 1 + 1 + 6 + 6 + 6 + 6$

15. If the sum of cubes of each digit of a number is equal to the number itself, the number is called **Armstrong number.**

$153 = 1^3 + 5^3 + 3^3 = 1 + 125 + 27$
$370 = 3^3 + 7^3 + 0^3 = 27 + 343$
$371 = 3^3 + 7^3 + 1^3 = 27 + 343 + 1$
$407 = 4^3 + 0^3 + 7^3 = 64 + 0 + 343$

16. The date 1888 AD requires the use of most Roman letters MDCCCLXXXVIII containing 13 letters, it was used on the entrance to the High Court of New South Wales stating its year of completion.

17. As far as the modern zero concept is concerned it is purely

Indian. As far back as 200BC, we get the earliest known reference to zero in ***Acharya Pingla's book Chanda-Shastra*** where he writes: "In Gayatri chanda, one pada has six letters; when the number is made half it becomes three. Remove one from it and make it half to get one. Remove one from it, thus get the zero."

गायत्रे षड़संख्यामधेर्ड पनीते द्वयङ्के अवशिष्ट स्त्रयस्तेषु ।
रूपमपनीय द्वयङ्कायः शून्यं स्थापत्यम् ॥

In mathematical words

½ {1/2 (6) -1} —1=0

Aryabhatta (476AD), a great Indian mathematician and astronomer, had devised a number system without zero though it was a positional system. He used the word **kha** (ख) for the positional system which in the later phase become a name for zero. Another renowned mathematician of India **Varahmihira** (505AD-587AD), in his famous book *Pancasiddhantika* written in 575AD, had many a times used the place value system containing zero which is believed to be a fairly proven fact for the use of zero for the first time.

It was the Indian mathematician **Brahmagupta** who in his book ***Brahma-Sputa-Siddhanta*** defined zero, He wrote this book the age of 30, i.e. in 628 AD where he writes

a—a = 0 (in modern symbol).

In the Khmer inscription of 683AD found at Trapeang Bay in Cambodia a heavy point appears for zero whereas in an inscription found at Sumatra in 683AD, small circle appears for zero.

18. Take any four-digit number whose all digits are not equal. Arrange this number in descending order and then reverse it. Subtract these two. Repeat the process. After a few steps you shall find 6174. This is called Kaprekar Constant. **D.R. Kaprekar** was a famous mathematician of India, born on Jan17, 1905.

Example:

Take a number	1234
Arrange in descending order	4321
Reverse it	– 1234
Subtract it	3087
Arrange in descending order	8730
Reverse it	– 0378
Subtract it	8352
Arrange in descending order	8532
Reverse it	– 2358

D R Kaprekar

19. Palindrom numbers are numbers which remain the same when read from left to right or vice-versa. It is also mentioned in **Mahavira's** book.

139×109=15151
333333666667×33=11000011000011
3003 × 37 = 111111
6006 × 37 = 222222
9009 × 37 = 333333
12012 × 37 = 444444
15015 × 37 = 555555
18018 × 37 = 666666
21021 × 37 =777777
24024 × 37 = 888888

27027 × 37 = 999999
152207 × 73 = 11111111
14287143 × 7 =10010001
12345679 × 9 =111111111
142857143 × 7 =1000000001
27994681 × 441 =12345654321

20. 40 when written 'forty' is the only number with letters in alphabetical order, while 'one' is the only one with letters in reverse order.

21. Pi (3.14159...) is a number that cannot be written as a fraction. The digits to the right of the π (pi) (3.141...) decimal point can keep going forever, and there is no pattern to these digits at all. The billionth digit of Pi is 9. 14th day of 3rd months is celebrated as Pi-Day.

22. The equation which contains five most important numbers in mathematics is;—

$$e^{i\pi} + 1 = 0$$

This famous equation was discovered by **Leonard Euler**.

Richard Feynman, the great American Physicst and Noble Laureate called this the most remarkable formula in mathematics.

23. Eleven (11) is the fourth number that stays the same when written upside down. The other three are 0, 1, and 8.

24. Between 9,999,900 to 10,000,000, there are only 9 prime numbers;—

9,999,901; 9,999,907; 9,999,929; 9,999,931; 9,999,937; 9,999,943; 9,999,971; 9,999,973 and 9,999,991.

But in the next 100 integers from 10,000,000 to10,000,100, there are only two prime numbers

10, 000, 019 and 10,000,079.

25. The largest known prime number is of 6,320,430 digits found by **Michael Shafer** in Dec 2003. It would need 1400 to 1500 pages to write it.

26. Two numbers are said to be Amicable Numbers if each is equal to the sum of all the exact divisors of the others except the number itself. 220 and 284 are Amicable Numbers. 220 has the exact divisors 1, 2, 4, 5, 10, 11, 20, 22, 44, 55 and 110 and whose sum is 284, and 284 has the exact divisors 1, 2, 4, 71 and 142 whose sum is 220.

The set of 220 and 284 was the first known set of amicable numbers. **Pythagoras** discovered the relationship and coined the term *Amicable* because he considered the numbers to be a symbol of friendship.

No other pairs were known until 1636 when **Fermat** discovered 17296 and 18416 as a second pair. This pair was actually discovered over three hundred years earlier by the Arab mathematician **al Banna**, but it was never known in the West until Fermat's finding.

Descartes discovered the third pair of 9,363,584 and 9,437,056.

There is an unauthenticated story related to Amicable numbers that a prince whose name from the standpoint of numerology was equivalent to 284 sought a bride whose name would represent 220 believing the fact that such a combination would guarantee for a happy marriage throughout the life. Here is the list of 13 Amicable numbers.

1	220	284
2	1184	1210
3	2620	2924
4	5020	5564
5	6232	6368
6	10744	10856

7	12285	14595
8	17296	18416
9	63020	76048
10	66928	66992
11	67095	71145
12	69615	87633
13	79750	88730

Leonard Euler in 1747 turned his attention to Amicable numbers and surprisingly, over a period of three years he produced 58 amicable numbers;

Leonard Euler

27. Thirty four (34) is the smallest number with the property that it and its neighbours have the same number of divisors. 34 has three divisors 1, 2 and 17 whereas its neighbour 35 too has three divisors 1, 5 and 7.

28. The Pythagorean believed that the even numbers (2, 4, 6…) were feminine and the odd numbers (3, 5, 7..) were masculine. They believed that odd numbers were superior to even numbers and that odd numbers were associated with good luck (except13) and the even number with bad luck.

The Pythagorean believed that the number 4 stood for justice because it was a perfect square (2^2). They believed that 5 stood for marriage because it was the sum of first even number and first odd number 5=2+3. The Pythagorean believed that the first natural number 1 was the divine, so for centuries after, 1 was believed to be somehow so different from the rest of numbers that it was neither called an odd number nor a natural number.

29. The existence of negative numbers in mathematics is a subject of discussion. There is no clear evidence of negative numbers being recognised in Babylonian, Greek or any other ancient civilization. In 17th century, negative numbers were used in book-keeping in India. Positive quantities denoted assets and negative quantities as debts. The Hindu astronomer, **Brahmagupta,** in 630 AD in his book dealing with mathematics shows a clear understanding of negative numbers. The earliest documented evidence of the use of negative numbers in European mathematics is the Ars Magna, published in 1545 by the Italian mathematician Girolamo Cardano.

Brahmagupta

30. The number 10 is used as a convenient base to count with, but the Gauls of ancient France, the Mayas of Central America, and other peoples used a base of 20. The Sumerians, the Babylonians, and others after them used a base of 60–convenient because 60 can be evenly divided by 2, 3, 4, 5, 6, 10, 12, 15, 20, and 30. The use of base 60 survives in the division of hours into minutes and minutes into seconds, and the division of the circle

into 360 (60 × 6) degrees.

31. **Ramanujan Number** – 1729, 4104, 13832 are called the Ramanujan-Hardy Number. 1729 is the least number which can be easily expressed as the sum of the cubes of two numbers in different ways.

$$1729 = 10^3 + 9^3$$
$$= 12^3 + 1^3$$

This was in fact the number of a taxi Prof Hardy had hired when he came to see the ailing Ramanujan in the hospital. When Prof Hardy asked Ramanujan that the taxi he had hired had a boring number; Ramanujan instantly said that this is the smallest number which can be expressed as the sum of cubes of two numbers in two different ways. Ramanujan was born on December 22, 1887 at Erode in Madras Presidency in a very poor family. He passed his primary education in 1897 scoring first in the district but could not pass his B.A. Prof Hardy was very much impressed when he got a letter from Ramanujan and invited him England. He was awarded the B.A. degree and he became the Fellow of the Royal Society. Prof Hardy wrote **"Ramanujan was my discovery. I did not invent him— like other great men, he invented himself— but I was the first really competent person who had the chance to see some of his work, and I can still remember with satisfaction that I could recognise at once what treasure I had found."** Ramanujan worked on partition numbers, hyper geometric series, and divergent series.

32. The beauty of some of the numbers has astonished the mathematician. One such number is 65359477124183.

65359477124183× 17=1111111111111111
65359477124183×34=2222222222222222

65359477124183×51=3333333333333333
65359477124183×68 =4444444444444444
65359477124183×85=5555555555555555
65359477124183×102=6666666666666666
65359477124183×119=7777777777777777
65359477124183×136=8888888888888888
65359477124183×153=9999999999999999

Multiply 65359477124183 by the multiple of 17's and get as many times the multiple of 1's.

33. Numbers of the type $F_n = 2^{2^n}+1$ for n =1, 2, 3… are called Fermat Numbers. Piere De Fermat wrote to Mersenne on December 25, 1640 that:

If I can determine the basic reason why 3, 5, 17, 257, 65537... are prime numbers, I feel that I would find very interesting results, for I have already found marvellous things (along these lines) which I will tell you about later.

Fermat thought that these numbers might be primes but it is not true for F_5. In 1732 Euler discovered that F_5 is divisible by 641.

$F_5 = 641 \times 6700417 = 4294967297$

Pierre de Fermat (1601-1665) was a brilliant and versatile French amateur mathematician. He is considered as the founder of probability theory along with Pascal. He was also considered to be the co-inventor of Analytic Geometry along with Rene Descartes

Fermat

34. Fibonacci Number:- 1, 1, 2, 3, 5, 8, 13... are known as the Fibonacci numbers. It is the set of numbers which after the second is the sum of the two previous ones.

1+1=2, 1+2=3, 2+3 =5 and so on. Mathematically, Fibonacci Series is defined as —

$$F_{n+1} = F_{n-1} + F_{n}, \text{ if } n > 1$$
$$F_0 = 0 \text{ and } F_1 = 1$$

Fibonacci or **Leonardo Pisano** was born in Pisa in 1170 but was educated in North Africa where his father was working. He had written many books of which *Liber Abaci, Liber quadratorum, Practica Geometriae* are still available. Liber Abaci, a book published in 1202, was based on Arithmetic and Algebra. This book introduced the European to the Hindu–Arabic Numerals and place value system. Fibonacci number is also discussed in Liber Abaci. He died in 1250 AD.

35. Liouville Number :- Any number of the form

$$\sum_{n=1}^{\infty} \frac{A}{10^{n!}}$$

where A is a constant, is called a Liouville Number

Liouville Joseph(1809-1882) was a French analyst and geometer. He was the first to prove the existence of transcendental numbers. In 1844, he proved that transcendental numbers existed by actually constructing examples. He also proved that such numbers cannot be the solution of polynomial equation with integer coefficients and hence are not algebraic.

36. **Pythagorean Triplet:-** Any set of positive integers satisfying the equation

$$x^2 + y^2 = z^2$$

is called Pythagorean numbers. Such numbers are given by $m^2 - n^2$, $2mn$, $m^2 + n^2$ where m and n are arbitrary positive integers with the condition that $m \neq n$. More over $n, \left(\frac{n^2-1}{2}\right), \left(\frac{n^2-1}{2}\right) +$ 1 are also the set of Pythagorean triplets. (3,4,5), (5,12,13), (6,8,10), (7,24,25), (8,15,17), (9,40,41), (11,60,61)... are examples of Pythagorean triplet.

Pythagoras of Samos (569BC-475BC), a great Greek mathematician, is famous for his theorem better known as Pythagoras Theorem which states — "the sum of the square of the lengths of the legs of a right triangle is equal to the square of the length of the hypotenuse."

He founded a school in Samos, his birthplace, which was called *Semicircle*. He made remarkable contribution in the field of music, mathematics and astronomy. Numbers were his personal friend.

He used to play with pebbles forming a group of different numbers. The pebbles made it possible for Pythagoreans to identify different shapes, the simplest being the two-dimensional figure—the triangle—and simplest three-dimensional figure was the tetrahedron.

Aristotle in his Metaphysics writes "They (the Pythagoreans) supposed the elements of numbers to be the elements of all things and the whole heaven to be a musical scale and a number... Evidently then these thinkers also consider that number as the principal, both as matter for things and as forming both their modification and their permanent states."

Pythagoras

37. **Lucas Numbers:**—A series of numbers
2, 1, 3, 4, 7, 11, 18, 29, 47, 76, 123, 199, 322, 521...
is called Lucas numbers

$$L_n = L_{n-1} + L_{n-2}$$

for n>1 is called Lucas series where $L_0 = 2$ and $L_1 = 1$.

Edouard Lucas (1842-1891) a French mathematician is best known for his result in number theory. He studied the Fibonacci sequence and found that if the Fibonacci rule of adding the last two to get the next is kept and the series is started from 2 and 1 instead of 0 and 1, a new series 2, 1 , 3, 4, 7, 11, 18,... is formed. This series is called Lucas series in his honour. Lucas is also known for the invention of the Tower of Hanoi Puzzle and other mathematical recreation.

38. The number 666 is called the **Beast number**. The direct reference of the number 666 can be found in the last book of Bible— Revelation in chapter 13 verse 18.

During the second world war Adolf Hitler was bestowed the title of Beast. If A=100, B = 101.... Z=125 then HITLER comes out to be 666.

H=107

I= 108

T=119

L=111

E=104

R=117

HITLER = 666

Let us check some characteristics of 666.

$$666 = 1^6 - 2^6 + 3^6$$

$$= 6 + 6 + 6 + 6^3 + 6^3 + 6^3$$

$$= 1 + 2 + 3 + 4 + 567 + 89$$

$= 123 + 456 + 78 + 9$
$= 9 + 87 + 6 + 543 + 21$
$= 2^2 + 3^2 + 5^2 + 7^2 + 11^2 + 13^2 + 17^2$

666 is equal to the sum of the digits of its 47th power and is also equal to the sum of its digits of its 51st power:

39. The sum of 7 consecutive prime numbers equals a prime number. The same pattern holds for 4 groups of 5 primes starting with number 5.

$$17 + 19 + 23 + 29 + 31 + 37 + 41 = 197$$
$$19 + 23 + 29 + 31 + 37 + 41 + 43 = 223$$
$$23 + 29 + 31 + 37 + 41 + 43 + 47 = 251$$
$$29 + 31 + 37 + 41 + 43 + 47 + 53 = 281$$
$$31 + 37 + 41 + 43 + 47 + 53 + 59 = 311$$

40. **3816547290** is the only base 10 number that—

- uses each of the 10 digits exactly once
- the leftmost k digits are evenly divisible by k.

In other words. If we take 2 digit from left say 38 then it will be divisibly by 2. If we take three digits from left i.e. 381, it will be divisibly by 3. The number 3816547290 has the unique property of being divisible by 10, 9, 8, 7, 6, 5, 4, 3, 2, 1 after striking off the right most digits successively \

3	is evenly divisible by 1
38	is evenly divisible by 2
381	is evenly divisible by 3
3816	is evenly divisible by 4
38165	is evenly divisible by 5
381654	is evenly divisible by 6
3816547	is evenly divisible by 7
38165472	is evenly divisible by 8
381654729	is evenly divisible by 9
3816547290	is evenly divisible by 10

41. Prime numbers can also work as Palindrome number. Here is the list of primes that are also Palindromes.

1878781 188888881
1879781 123494321
1880881 120343021
1881881 323232323

The list is endless.

42. The number 9 is the only number that when it is multiplied by any other number (no matter how large that number is) and the individual numbers of the answer are added together until you only have a single number the result is always 9.

$9 \times 1 = 9$
$9 \times 2 = 18 = 1 + 8 = 9$
$9 \times 12 = 108 = 1 + 0 + 8 = 9$
$9 \times 2456 = 22104 = 2 + 2 + 1 + 0 + 4 = 9$
$9 \times 3769287 = 33923583 = 3 + 3 + 9 + 2 + 3 + 5 + 8 + 3 = 36$
$= 3 + 6 = 9$

Moreover, $\frac{1}{9} = 0.111...$ $\frac{2}{9} = 0.222...$ $\frac{3}{9} = 0.333...$ $\frac{4}{9} = 0.444...$

$\frac{5}{9} = 0.555...$ $\frac{6}{9} = 0.666...$ $\frac{7}{9} = 0.777...$ $\frac{8}{9} = 0.888...$

43. Perhaps the numeral seven is considered the most mystical of all numbers because it is the one number which cannot be divided evenly into the circle. Threr is always a remainder when 360 is divided by 7 see the chart.

CIRCLE	DIVISOR	DIVIDEND
360	1	360
360	2	180
360	3	120

360	4	90
360	5	72
360	6	60
360	7	51.428571
360	8	45
360	9	40

44. The first record of Indian use of zero can be found on the inscription on a stone tablet, at the town of Gwalior of king Jayavardhana II, written around 876AD, where the number 270 and 50 are written as we write today; the only difference is that the 0 is smaller and slightly raised. The description is as follows:

They have planted a garden 187 by 270 hand which would produce enough flowers to allow 50 garlands per day to be given to the local temple.

45. Triangular number can be seen in Pascal's triangle. Look at the Pascal's Triangle and you will find that the third diagonal is all triangular numbers.

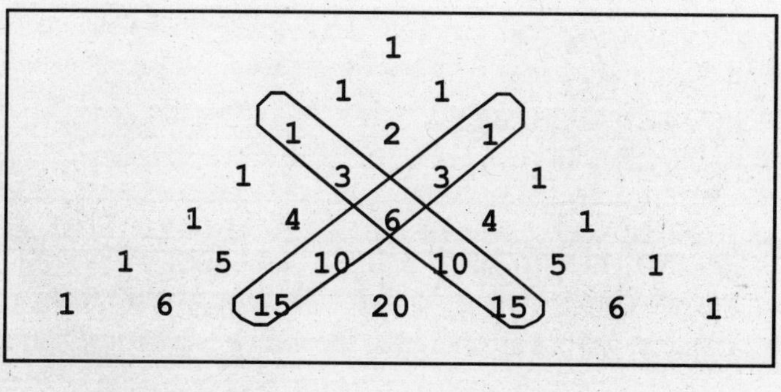

Here, 1, 3, 6, 10, 15… are the example of triangular number.

46. Triangular number: Number when arranged in dots makes a triangular shape Example: 1, 3, 6, 10, 15

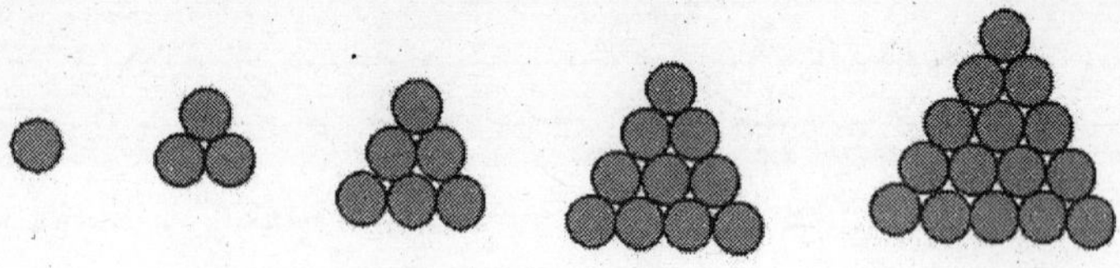

1 3 6 10 15

The sum of consecutive triangular number is a square number

$T_1 = 1^2$

$T_2 + T_2 = 1 + 3 = 4 = 2^2$

$T_2 + T_3 = 3 + 6 = 9 = 3^2$

$T_3 + T_4 = 6 + 10 = 16 = 4^2$

$T_4 + T_5 = 10 + 15 = 25 = 5^2$

$T_5 + T_6 = 15 + 21 = 36 = 6^2$

$T_6 + T_7 = 21 + 28 = 49 = 7^2$

$T_7 + T_8 = 28 + 36 = 64 = 8^2$

$T_8 + T_9 = 36 + 45 = 81 + 9^2$

$T_9 + T_{10} = 45 + 55 = 100 = 10^2$

The Triangular Number has another very fascinating relation with a square number. It you multiply any triangular number by 8 and add 1, you will get a square number

$1 \times 8 + 1 = 9 = 3^2$

$3 \times 8 + 1 = 25 = 5^2$

$6 \times 8 + 1 = 49 = 7^2$

$10 \times 8 + 1 = 81 = 9^2$

$15 \times 8 + 1 = 121 = 11^2$

One particular triangular number that Pythagoras especially liked was the number ten. It was called a Tetractys, meaning a set of four things, a word attributed to the Greek Mathematician and astronomer Theon.

Theon (335-405) was the father of Hypatia (a great women mathematician). He edited and arranged Euclid's Elements and Ptolemy's Handy Tables.

Theon

47. Every Perfect Number except 6 can be written as a sum of the cubes of an unbroken sequence of consecutive odd integers.

$28 = 1^3 + 3^3$

$196 = 1^3 + 3^3 + 5^3 + 7^3$

$8128 = 1^3 + 3^3 + 5^3 + 7^3 + 9^3 + 11^3 + 13^3 + 15^3$

____

____

48. Most important thing about the perfect number is that the single digit across sum of every perfect number except 6 is 1.

$28 = 2 + 8 = 10 \Rightarrow 1 + 0 = 1$

$496 = 4 + 9 + 6 = 19 \Rightarrow 1+9 = 10 \Rightarrow 1 + 0 = 1$

$8128 = 8 + 1 + 2 + 8 = 19 \Rightarrow 1 + 9 = 10 \Rightarrow 1 + 0 = 1$

49. The largest numbers known so far are googol and googolplex.

1 googol = 10^{100}

1 googolplex = 10^{googol}

The name googol and googolplex were coined in the late 1930s by the 9-year-old child Milton Sirotta, nephew of the American mathematician Edward Kasner(1878-1955) when he was asked to come up with a name for a very large number.

50. Any two consecutive primes which differ by 2 are known as twin primes. For example—(3,5) (5,7) (11,13) (17,19) (41,43)

Interval	1-100	100-200	200-300	300-400	400-500	500-600
Twin primes	7	7	4	2	3	3

The largest pair of twin primes known so far is 140, 737, 488, 353, 700 ± 1

51. The great scientist Albert Einstein devised a formula for generating primes. He visualised that

$$2^2 + 1$$

$$2^{2^{2}} + 1$$

$$2^{2^{2^{2}}} + 1 \text{ etc.}$$

are all primes.

Gauss was so much impressed by this that he said –There have been only three epoch-making mathematicians in history-Archimedes, Newton and Einstein.

But later it was found that this is not a complete criterion.

$2^{2^{2^{2}}} + 1 = 2^{65536} + 1$ is divisible by 825753601

and therefore not a prime.

52. In Binary number system, the odd number ends with 1 and the even number always ends in 0.

2	14	Remainder
2	7	0
2	3	1
2	1	1

$(14)_{10} = (1110)_2$

2	17	Remainder
2	8	1
2	4	0
2	2	0
	1	0

$(17)_{10} = (10001)_2$

The binary arithmetic, in which only two symbols 0 and 1 are used to represent all numbers was first given by Leibniz in 1679.

53. Fibonacci (1170-1250) in his book Liber Abaci published in 1202 introduced the Hindu-Arabic number popular in Europe. Fibonacci writes in Liber-abaci that with the help of the nine Indian figures 9,8,7,6,5,4,3,2,1 and 0 (zero) any number may be written.

54. There are 20 digits in the expansion of 2^{64} but surprisingly the base 2 disappears.

$$2^{64}=1,84,46,74,40,73,70,95,51,616$$

55. Four is the smallest number of colours sufficient to colour

all planar maps. The four colour problem dates to 1852, when **Francis Guthrie**, while trying to colour the map of counties of England, noticed that four colours are sufficient to map any planar maps so that no two adjacent countries have the same colour.

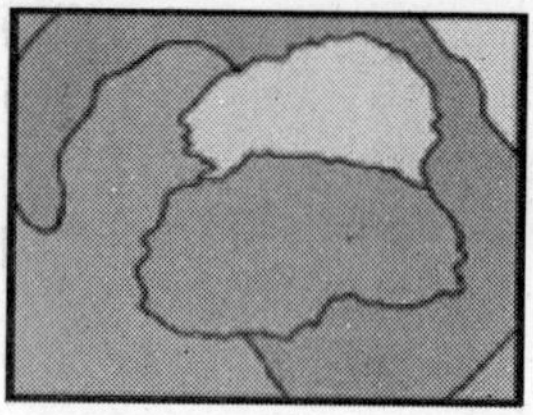

56. Four is the smallest honest number. Honest numbers are numbers **n** which can be described by using **n** letters of English alphabet. 4 in English is written as FOUR, which has exactly four letters.

57. The opposite side of a die cube always adds up as 7.

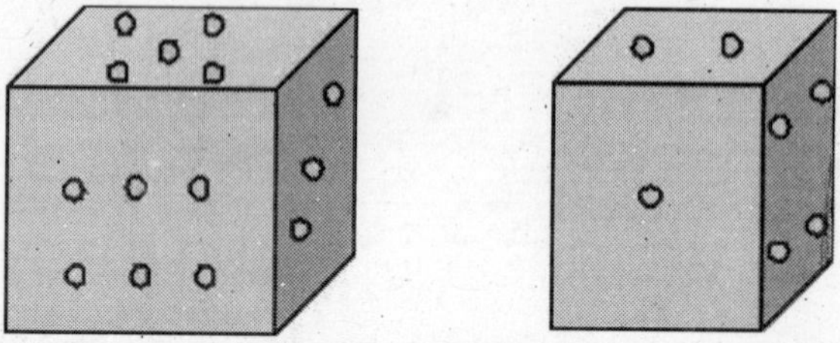

Game on Dice: Take any tow unbiased die. Throw it and note down the outcome. Lift the dice and note down the digit at the bottom. Now Do these operation and you will always get 49.

49 = (upper digit on 1st dice) × (upper digit on 2nd dice) + (upper digit on 1st dice) × (bottom digit on 2nd dice) + (upper digit on 2nd dice) × (bottom digit on 1st dice) + (Lower digit on 1st dice) × (lower digit on 2nd dice)

The digit on opposite face of 5 is 2 and 5 + 2 = 7.

The digit on opposite face of 6 is 1 and 6 + 1 = 7.

The digit on opposite face of 4 is 3 and 4 + 3 = 7.

Example: Suppose you get 3 and 6 on the dice, its bottom digit will be 4 and 1. Now do the operation 3 × 6 +3 × 1 + 4 × 6 +

4 × 1 = 18 + 3 + 24 + 4 = 49.

58. The digital sum of a prime pair product is 8.
5 × 7 = 35 and 3 + 5 = 8
11 × 13 =143 and 1+4+3 = 8
17 × 19 = 323 and 3+2+3 = 8
29 × 31 = 899 and 8+ 9 +9 = 26 and 2+6 = 8.

59. A partition of a number **n** is a decomposition of n into a sum of positive integers. Number of all possible partition, disregarding order, for a number **n** we denote by p(n). The formula for partitions is, probably, the best known portion of the **Ramanujan's** work. Thus, for example, 5 can be portioned in 7 ways.

5 = 1 + 1 + 1 + 1 + 1
5 = 1 + 1 + 1 + 2
5 = 1 + 1 + 3
5 = 1 + 4
5 = 2 + 3
5 = 2 + 2 + 1
5 = 5

P(n)	**Number of partition**
P(0)	1
P(1)	1
P(2)	2
P(3)	3
P(4)	5
P(5)	7
P(6)	11
P(7)	15
P(8)	22
P(9)	30
P(10)	42

60. Mersenne Numbers are integers of the form $M_n=2^n-1$. These numbers are named after the French monk Mersenne because he encouraged many mathematicians to study them and incorrectly conjectured that the Mersenne numbers were prime for n = 2, 3, 5, 7, 13, 17, 19, 31, 67, and 257; and all the other Mersennes with n < 257 were composite.

Marin Mersenne (1588-1648) was a Monk in a Church in France. He investigated the prime numbers and worked on finding formulae that would represent all the primes. Though he failed to do so but in the process he conjectured that

in $M_n = 2^n - 1$, if n = 2, 3, 5, 7, 13, 17, 19, 31, 67, and 257, the Mersenne Numbers would be Prime.

Mersenne

In December 2005 a Mersenne prime called M30402457 has been discovered by the mathematicians of Central Missouri State University Britain which contains 91 lac digits.

61. 17 is the only prime number that is the sum of four consecutive primes.

$$17 = 2 + 3 + 5 + 7.$$

It is the smallest number that can be written as $A^2 + B^3$ in two different ways.

$$17 = 3^2 + 2^3$$
$$= 4^2 + 1^3.$$

62. 36 is the smallest number besides 1 which is both a square and a triangular number.

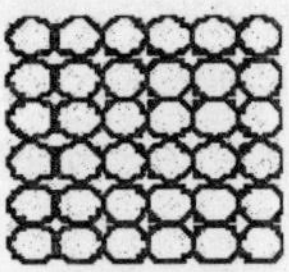

Square Number

Triangular Number

63. The smallest number whose English name contains all five vowels exactly once is 206.

206 = Tw<u>o</u> H<u>u</u>ndr<u>e</u>d <u>A</u>nd S<u>i</u>x

↓ ↓ ↓ ↓ ↓

1 2 3 4 5

64. 353 is the smallest number whose fourth power can be written as the sum of 4 fourth powers.

$$353^4 = 30^4 + 120^4 + 272^4 + 315^4$$

65. The symbol φ(phi) is called the Golden number.

$$\varphi = \frac{1+\sqrt{5}}{2} = 1.6180339887...$$

Mathematician **Mark Barr** proposed using the first letter in the name of Greek sculptor Phidias, phi, to symbolise the golden ratio. Usually, the lowercase form (φ) is used. Sometimes, the uppercase form (Φ) is used for the reciprocal golden ratio, 1/φ. The recent research in year 2010; on Golden ratio has helped researchers from the University of California to discover that the secret of female facial beauty lies in golden ratio. They have found that the distance between her eyes and her mouth are key factors in determining how attractive she is to others. Pamella Pallett and Stephen Link of UC San Diego and Kang Lee of the University of Toronto successfully identified the optimal relation between eyes, the mouth and edge of the face for individual beauty.

They found two Golden ratios, one for length and one for the width. Female faces were judged more attractive when the vertical distance between their eyes and mouth was approximately 36% of the face's length, and the horizontal distance between their eyes were 46% of the face's width. These proportions correspond with those of an average face.

66. A number is called **happy number** if the sum of square of the digits of any number on repentance gives 1. e.g. 19 ($1^2+9^2 = 82 = 8^2+2^2 = 68 = 6^2+8^2 = 100 = 1^2+0^2+0^2 = 1$).

On the other hand, if the sum of square of the digits of any number leads to a recurring cycle of numbers it is called **sad number.**

e.g. 12 ($1^2+2^2=5$, $5^2=25$, $2^2+5^2=29$, $2^2+9^2=85$, $8^2+5^2=89$, $8^2+9^2=145$, $1^2+4^2+5^2=42$, $4^2+2^2=20$, $2^2+0^2=4$, 42=16, $1^2+6^2=37$, $3^2+7^2=58$, $5^2+8^2=89$.....). Here the repentance starts from 89, hence 12 is a sad number.

67. Canadian Mathematician Michael Blake has created a piece of music based on mathematical constant pi to celebrate Pi-Day the 14th day of 3rd Month [$\pi - 3.14$]

The 36 years old composed a musical interpretation of Pi to 31 decimal places in 2011. He also set the piece to a tempo of 157 beats per minutes i.e. $157 = \frac{314}{2}$

□

2

Mathematicians

68. Alice's Adventure in Wonderland the best loved children's book was written by a professional mathematician Charles Lutwidge Dodgson whose pen-name was Lewis Caroll.

Lewis Caroll

69. **William Roman Hamilton,** an Irish mathematician, devised vector (a quantity having magnitude as well as direction) to solve a geometric calculation.

William Roman Hamilton

70. **Euclid** (330-275 BC), a great Greek mathematician better known as the Father of Geometry published 13 volumes of his book Elements dealing only with Geometry and Number Theory.

A page from Euclid's Elements.

71. **Simon Stevin** in 1585 for the first time in a small pamphlet The Tenth advocated the use of decimal fraction.

Simon Stevin

72. **Christine Franklin** (1847-1930), a woman mathematician who was doing research on 'The ALGEBRA of Logic' got her PhD after 44 years of submitting her thesis. She had completed her thesis in 1882 and Johns Hopkins University awarded her a PhD in 1929.

Christine Ladd-Franklin

73. **Leonardo da Vinci** was called *Ten men in one* because he was a painter, inventor, lute player, sculpture, military engineer, scientific observer, anatomist, architect, town planner, and designer.

Leonardo da Vinci

74. While at school **Ramanujan** studied a book **"A synopsis of Elementary Results in Pure and Applied Mathematics"** written by George Shoobridge Carr, containing 5000 equations in algebras, calculas, trigonometry and analytical geometry, which Ramanujan solved without taking any help.

Ramanujan

75. **Charles Reason** (1814-1893) was probably the first African American to receive the faculty position in mathematics at a predominantly white institution-Central College in Cortland Country, New York.

Charles Reason

76. **Abul Wafa** was born on 10 June 940 in Khorasan region in Iran. He is known for the first time use of tan function and compiling tables of sines and tangent at 15' intervals, he devised a

new method of calculating sine table which was correct up to 8 decimal places.

77. **Jacques Bernoulli** studied the logarithmic spiral r=a? and was so much fascinated that he wanted the curve to be engraved upon his tombstone with the words *"Eden mutata resurgo"* meaning I shall arise the same though changed.

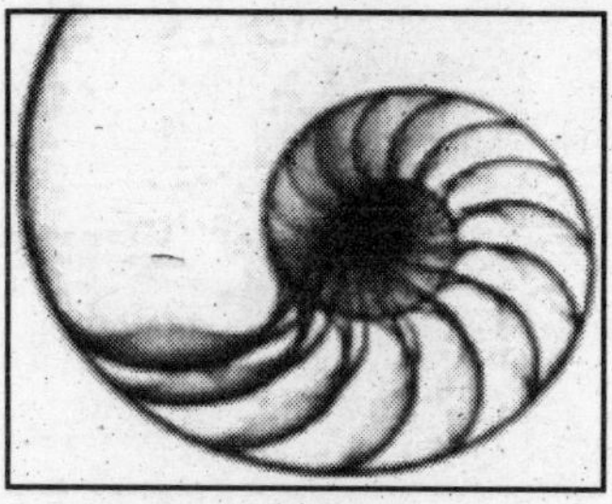

Logarithmic shape in sea shell

78. The log invented by **Napier** is different from what we use today. In Napier's logarithm, the value of log $1 \neq 0$ while in modern logarithm we take $\log 1 = 0$. It was Briggs who was responsible for this change though he had given credit for this to Napier.

John Napier

79. Geminus is said to have divided mathematics into two parts Pure and Applied. Pure mathematics included Arithmetic and Geometry whereas Applied Mathematics contained with mechanics, Astronomy, Aptics, Geodesy, Canonics and Logistic.

80. **Christopher Wren** (20 Oct 1632 - 25 Feb 1723) was

more famous as an architect of a church than as a mathematician. He used to be accorded responsibility for rebuilding 15 churches in the city of London after the great fire in 1666. He was a founder of Royal Society.

Christopher Wren

81. A French computer scientist **Fibrice Bellard** claims to have set a record in the calculation of mathematical constant pi. According to Fibrice Bellard, he calculated the value of pi to nearly 2.7 trillion digits, some 123 billion more than the previous record. Bellard used a desktop computer to perform the calculation, taking a total of 131 days to complete and also check the result reports BBC. This version of pi takes over a terabytes of hard disk space to store. Previous records were established using supercomputers, but Bellard claims his method is 20 times more efficient. The previous record of about 2.6 trillion digits set in August 2009 by Daisuke Takahashi at the University of Tsukuba in Japan, took just 29 hours. However that work employed a supercomputer 2000 times faster and 1000 times more costly than the desktop Bellard employed. It is difficult to estimate how long the currently determined pi will take to count, but if you count one number per second it would take more than 49000 years.

82. **Peter Ramus** (a great mathematician) belonged to a very impoverished family. His grandfather was a charcoal burner and his father was a peasant. After struggling with many odds he got

employment as a servant to a rich student in the College de Navarre at Paris and made his way to the master's degree. He became famous for his assertion that whatever Aristotle told was false.

Peter Ramus

83. **Arthur Cayley** (1821-1895) was a prominent mathematician. He had three papers published in the newly founded Cambridge Mathematical Journal when he was an undergraduate. He had published over 900 papers and notes covering every aspect of modern mathematics. He practiced for14 years as a lawyer. During these 14 years he published about 250 mathematical papers.

Arthur Cayley

84. **Thales of Miletus** was one of the seven wise men of

Greece. He was a Mathematician, Astronomer, Artist, Statesman, Engineer, Philosopher and man of business.

Thales (624-547BC)

Thales believed that the Earth floats on water and all things come to be from water. For him, the Earth was a flat disc floating on an infinite ocean. It has also been claimed that Thales explained earthquakes from the fact that the Earth floats on water. Again the importance of Thales idea is that he is the first recorded person who tried to explain such phenomena by rational rather than by supernatural means. Thales put geometry on a logical footing and was well aware of the notion of proving a geometrical theorem. Thales is credited with five theorems of elementary geometry:

i. A circle is bisected by any diameter.
ii. The base angles of an isosceles triangle are equal.
iii. The angles between two intersecting straight lines are equal.
iv. Two triangles are congruent if they have two angles and one side equal.
v. An angle in a semicircle is a right angle.

85. **Saayana** (1315-1387AD), a minister of King Bukka I of the Vijaynagar Empire, mentions about the speed of light which comes precisely to 186413.22 miles/second that is very near to the present value 186300 miles per second. While giving his commentary on 4th verse of the hymn of Rigveda 1.50 he writes:

तथा च स्मर्यते योजनानां सहस्रे द्वे
द्वे शते द्वे च योजने एकेन निमिषार्धेन
क्रममाण नमोऽस्तु ते इति।

O Sun, bow to you, you who traverse 2202 yojanas in half a nimisha.

One yojna is equal to 9 miles, 110 yards = 9.06625 miles and half nimisha = 0.10666-second. Using these values, the velocity of light comes out to 186413.22 miles per second which shows that Indian knowledge was far better than that of the Western.

86. **Leonard Euler** (1707-1783) is considered as one of the greatest mathematician so far produced. He completed his Bachelor's degree at the age of 15 and his Master's degree at the age of 16 and at the age of 18 he published his first mathematical paper. Tragically, at the end of 28 he lost the sight of his right eye and around 60 he became completely blind, but he did not take rest. While blind he published over 400 mathematical papers, most of which he dictated to his servant untrained in mathematics. The Swiss mathematician Leonardo Euler is credited for creating the mathematical branch Topology in 1730. He is credited to have introduced the symbol i (iota), e (exponential), $\sum$ (summation) etc, to the world.

Euler

87. **Edmund Halley,** the eminent astronomer and Newton's best friend, bore the printing cost of Newton's Principia and also did its proof reading. The original name of this book is Mathematical Principles of Natural Philosophy which was published in 1687.

Edmund Halley

88. **Jean Robert Argand** (1768-1822) was one of the first to publish an account of graphical representation of Complex Number in 1806.

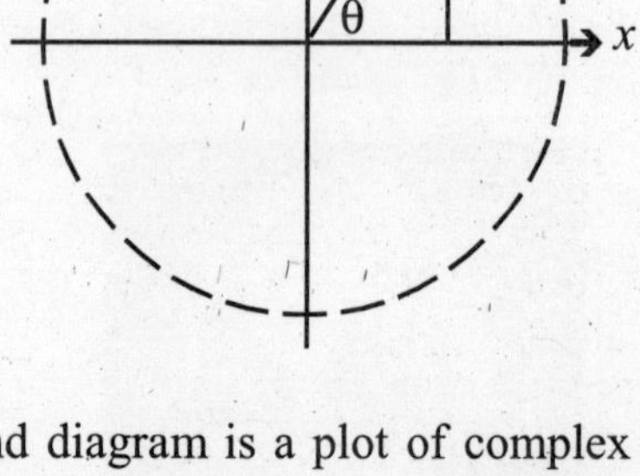

An Argand diagram is a plot of complex number as point z = x + iy. In the complex plane, the x-axis is taken as the real axis and y-axis as the imaginary axis. In the plot above the dashed circle represents the complex modulus $|z|$ of z and the angle θ represents its complex argument.

Any number real or imaginary of the form of z = x+iy is

called complex number where **a** and **b** are real numbers and $i^2 = -1$.

2+3i, 3+4i, 3-2i, -2-3i all are the example of complex number.

John Argand

In order to represent a complex number on a graph, the horizontal axis is considered as the axis of real numbers whereas the vertical line through 0 is considered as the axis of imaginary numbers.

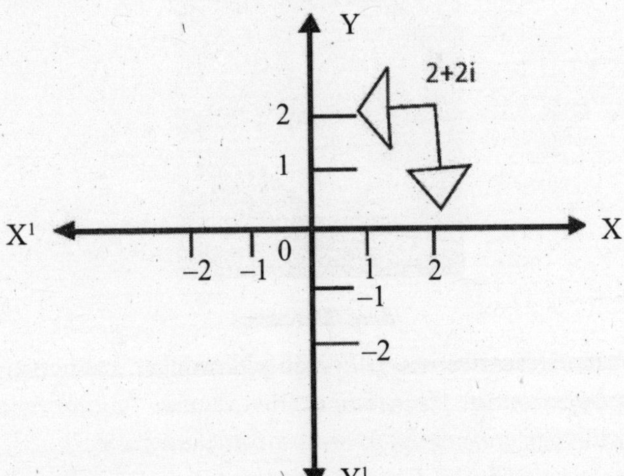

Suppose we have the following equation to solve

$$X^2 + 1 = 0$$

There is no number whose square is equal to −1, in simple

words we can say that we can't find the solution of x = $\sqrt{-1}$ until we define a number whose square is equal to $\sqrt{-1}$. Euler defined a number i (iota) whose square is equal to $\sqrt{-1}$. This iota is called an imaginary number. The name imaginary was given by Rene Descartes in 1637.

$i^2 = -1$ $\quad$ $i^3 = -i$ $\quad$ $i^4 = 1$ ....

89. **Rene Descartes,** the great French Mathematician of the 17th century, liked to lie in bed and think. One day when resting in bed he solved the problem of describing the position of a point in a plane. His method was a development of the older ideas of latitude and longitude. In honour of Descartes, the system used for describing the position of a point is also known as Cartesian system.

Rene Descartes

Rene Descartes was a French philosopher, mathematician, lawyer, and scientist. He is famous for his quote, "I think therefore I am" which he considered to be a self-evident truth.

Descartes broke from the traditional ways of thinking to establish a new system, which is based on logical reasoning. This deductive way of thinking encompassed all that he did. He is considered to be the father of the modern school of Philosophy and the first of the modern school of mathematics. Of his many writings are Discours de la Methode (1637)(Discourse on Method), written in French, and Regulae ad Dirctionem Ingenii (Rules for the Direction of the Mind). In these publications, Descartes outlined steps implementing mathematical procedures in the reasoning process.

90. In 1889, the Italian mathematician **Giuseppe Peano** formulated a set of 5 axiom that could be used to deduce the theorem of arithmetic.

a. One is a number
b. If x is a numbers, the successors of x is also a number.
c. One is not the predecessor of any number.
d. If two numbers have equal successors they are equal.
e. If a set of numbers contains the number one and it contains all the successors of its members, then the set contains all the numbers.

From these simple axioms and beginning with the first natural number one, Peano could define all natural numbers.

Peano

91. **Aryabhatta** I (476-550) in his Aryabhattia used the word jya for sine. The Arabs phonetically derived the word jiba from the Aryabhatta's text. When it was written in Arabic it was written as jb due to common practice of omitting the vowels. The later writers thinking it a meaningless word translated it into jaib meaning

pocket. When this Arabic text was translated into Latin it got the name sinus. The word sin is derived from this sinus.

92. **Colin Maclaurin** (1698-1746) died in 1746 but one of his mathematical papers was published in 1996. He was a Scottish mathematician who published the first systematic exposition of Newton's method written as a reply to Berkeley's attack on calculus for its lack of rigorous foundation.

About Maclawrin

93. **Isaac Barrow** (1630-1677) was a mathematics teacher and also a mathematician who was best known as a teacher of Newton. He resigned from his chair at the age of 39 to make way for Newton's appointment in his place as Professor of mathematics at Cambridge.

Issac Barrow

94. **George Boole** (1815-1864), who had given the rules for the new branch of Algebra called BOOLEAN ALGEBRA, was a school teacher. His most important work 'The Laws of Thought' was published in 1854.

George Boole

95. **Pierre Simon Marquis De Laplace** (1749-1827) taught Napoleon Bonaparte for a year when he was a professor in Paris.

Laplace

96. **Gert Mittring** a 38 year old German computer scientist broke the world record in mental calculation. He took 11.80 seconds to work out a 13th root of a 100 digit number.

97. **William Shanks** in 1853 calculated the value of π correct to 707 decimal places, and all by hand. In 1945, 92 years later it was found that the last 180 digits were wrong.

98. **Taratgalia** (? 1500-1557) was the first man to apply mathematics in the artillery. He abandoned the concept of a piecewise linear trajectory given by Aristotle and developed a new concept while giving the rules for the art of bombardier. In his book

Nova Scientia he described the new ballistic method and instruments including the first firing tables.

Taratgalia

99. The Indian text on which al-Khwarizmi based his treatise was one which had been given to the court in Baghdad around 770 as a gift from an Indian political mission. There are two versions of al-Khwarizmi work which he wrote in Arabic but both are lost. In the tenth century, al-Majriti made a critical revision of the shorter version and this was translated into Latin by Adelard of Bath. There is also a Latin version of the longer version and both these Latin works have survived. The main topics covered by al-Khwarizmi in the Sindhind zij are calendars; calculating true positions of the sun, moon aand planets, tables of sines and tangents; spherical astronomy; astrological tables; parallax and eclipse calculations; and visibility of the moon.

Al-Khwarizmi Algebra

100. **Archimedes,** an all time genius of mathematics, was bestowed by different titles such as **the God of Mathematics, the Homer of Geometry, and a genius more divine than human.** Archimedes is the first person known to have discussed extremely

large numbers. In his essays *The Sand Reckoner,* he built a system on myriad, the name for 10000. He called the numbers up to myriad myriads (10 000 000). Archimedes had invented a system of arithmetic by which he could express in language enormously large numbers in fact all numbers up to that which we would write as 1 followed by 80000 million million ciphers. He was killed by a Roman soldier while he was busy solving a sum on sand.

Archimedes

101. **Girolamo Cardano** (1501-1576) was the first to publish general solutions to cubic and quartic equations, and the first to publish the use of complex numbers in calculations. Cardano introduced binomial coefficients and the binomial theorem, and introduced and solved the geometric hypocyloid problem, as well as other geometric theorems.

Girolamo Cardano

102. **James Taylor** an Englishman, in 1816 translated Bhaskara's ***Lilavati*** into English. **Bhaskaracharya** was an Indian mathematician born in 1114 AD in Bijapur district in Karnatka. There is an unauthenticated story about Lilawati, the daughter of

Bhaskara II. It is said that astrological calculations had shown Bhaskara that, if the marriage of his daughter was not performed at a particular hour on that particular day, she would become a widow. And he had bought the water-clock to ensure that he would know the right time. Lilavati looked entranced at the water-clock her father had brought home. Its movements were fascinating. She had a slight feeling of guilt, for her father had told her never to enter that room. But because it was forbidden territory, its exploration gave her a sense of adventure. And she continued looking at the clock. Then came disaster, though she was never to know about it. A tiny pearl slipped out of her nose-ring and fell into the clock. This made Bhaskaracharya predict wrongly; the marriage was solemnised and a few days later she became widow.

Lilawati, the mathematical book, is a tribute of Bhaskaracharya to her daughter.

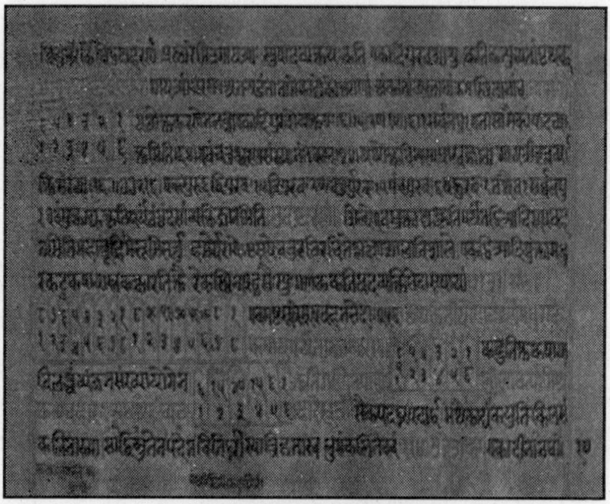

Lilawati of Bhaskaracharya

Let us see the poetic beauty of Lilawati
A row of pearls mislaid
One-Sixth fell to the floor.
One-Fifth upon the bed.

The young woman saved one third of them.
One tenth were caught by her lover.
If six pearls remained upon the string.
How many pearls was there altogether?
Ans = 16

103. **Vieta,** the first mathematician to employ vowel for unknowns and consonants for the known was a lawyer by profession who later became a member of parliament of Bretagne.

104. **Johann Mülles** a German mathematician of the 15th century in 1464 wrote De Triangles Omnimodis Libri, a first book solely devoted to mathematics.

105. **Archimedes** in his book 'Measurement of the circle' had shown that the value of π lies between $3\frac{10}{71}$ and $3\frac{1}{7}$ which he obtained by circumscribing and inscribing a circle with regular polygon having 96 sides. Probably this is the reason we take $\pi = 3\frac{1}{7} = \frac{22}{7}$

106. In the eighteenth century, the French naturalist **Comte de Buffon** showed that you could estimate π by experiment using a needle and a shove-halfpenny board, which consists of a grid of parallel lines.

Buffon

Suppose the lines are distant **a** from one another, and the needle has length **l** with **l < a** (so the needle can only cross one line at a time). Then you can conduct an experiment by dropping the needle randomly onto the board and counting the number of times the needle crosses any line. Suppose you do this n times, where n is large, and c of those times the needle crosses a line then—

$$\pi \simeq \frac{2nl}{ca}$$

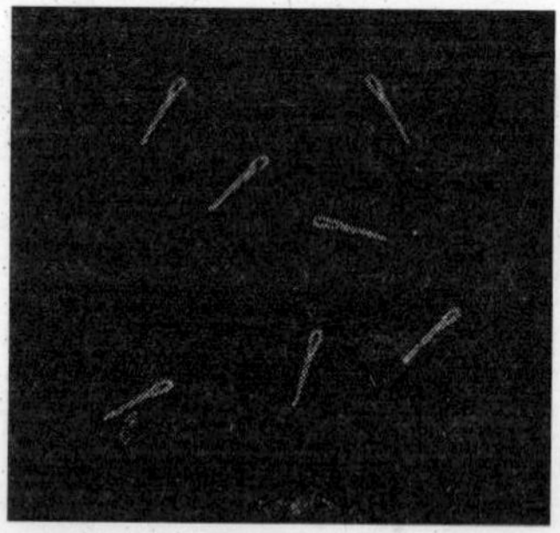

107. **Euclid** is the most successful textbook writer of all time. His *Elements*, written around 300 B.C., has gone through more than 1,000 editions since the invention of printing. *Elements*, the Euclid treatise which is sometimes said next to the Bible was first printed in 1482 and since then more than 1000 editions have been published.

Euclid

108. **Gallileo** was the first to realise that if air resistance is neglected the path of a projectile is a parabola. He was a famous Italian Physicst. He was born on 15th Feb, 1564 in Pisa and died on 8th Jan 1642 in Aratri in Italy.

Gallileo

109. The Indian mathematician **Aryabhatta** was born in Patna in 476AD. He was the first person to say that the earth is spherical and that it resolves around the sun.

The following formulae on mathematical series are due to him.

$1+2+3+.....+n = n(n+1)/2$

$1^2+2^2+.....+n^2 = n(n+1)(2n+1)/6$

$1^3+2^3+....+n^3 = \{n(n+1)l2\}^2$

110. **Hipparchus** (190–120 BC), a Greek mathematician was the first to prepare trigonometric tables. He is also credited for the invention of Trigonometry. He was the first whose quantitative and accurate models for the motion of the Sun and Moon survive.

Hipparchus

111. Here is a list of some mathematician who were bestowed titles.

Name of the mathematician	Title
Galileo	Father of Dynamics
Diophantus	Father of Algebra
Monge	Father of Descriptive geometry
Appolonius of Perga	The great Geometer
Leobachevsky	Copernicus of Geometry
Archimedes	The God of Mathematics
Newton	Father of Calculus
Peacock	Euclid of Algebra
Democritus	Laughing Philosopher
Euclid	Father of Geometry
Brahamagupta	Ganak chakra churamani
La Roche	Master of Cipher
Gauss	Prince of Mathematics

112. The life of **Diophantus** is known due to this problem written about 250AD.

His boyhood lasted 1/6 of his life, his beard grew after 1/12 more, after 1/7 more he married and his son was born after 5 years, the son died half of his father life and the father died 4 years after his son.

This problem tells that Diophantus died at the age of 84.

Diophantus

113. The Indian mathematician **Brahmagupta (598-668)** was the first to define zero in his book *Brahma-Sputa-Siddhanta* written at the age of 30 in 628 AD where he writes

$$a - a = 0$$

He was the head of astronomical Observatory at Ujjain and during his tenure he wrote four lext on mathematics and astronomy. In his Famous book Brahmasphuta Siddhanta; he mentioned about operation on Zero. In modern Symbol, his argument can be written as

$$a + o = a \quad a - a = o \quad o - a = -a$$

Brahmagupta

114. **Nicolaus Copernicus** (1473-1543), a Polish astronomer, proposed that the sun is at the centre of the Universe, with the earth and other planets revolving around it. The first print copy of his theory called De Revolutionibus Orbium Coelestium Librivi (Six books concerning the Revolutions of Heavenly orbs) was given to him on his deathbed. According to the legend, Copernicus awoke from his coma, looked at his book and died peacefully.

Copernicus

115. **Srinivas Ramanujan** was the first Indian as well as the first Asian to be elected to the Royal Society of London.

Ramanujan

116. **Theano,** a Greek women mathematician, was one of Pythagoras's pupils as a young girl. However, she soon married him. While they were married they started the first co-educational university. This school was based on equality between the sexes. She took over the school after Pythagoras died in a fire. She and her two daughters ran and taught at the school. Theano discovered her own rule and worked on a formula to derive the golden rectangle. The information that she gave to mathematics allowed Pascal to make his triangle and Fibonacci to derive his series.

Theano

117. The Indian mathematician and Astronomer **Aryabhatta** was the first to find the value of pi (π) correctly upto 4 decimal places.

चतुरधिकं शतमष्दगुणं द्वाषष्टिस्तथा सहस्राणाम् ।
अयुतद्वयविष्कम्भस्यासन्नो वृतपरिणाहः ॥

Add 4 to 100, multiply by 8 and add to 62000; this is apporoximately the circumference of a circle whose diameter is 20000.

In other words, if the diameter of a circle is 20000 and its circumference would be 62832 then

$$\pi = \frac{62832}{20000} = 3.1416$$

Aryabhata

Aryabhata's value of π is a very close approximation to the modern value and the most accurate among those of the ancients. There are reasons to believe that Aryabhata devised a particular method for finding this value. It is shown with sufficient grounds that Aryabhatta himself used it, and several later Indian mathematicians and even the Arabs adopted it. The conjecture that Aryabhata's value of π is of Greek origin is critically examined and is found to be without foundation. Aryabhata discovered this value independently and also realised that π is an irrational number. He had the Indian background, no doubt, but excelled all his predecessors in evaluating π. Thus the credit of discovering this exact value of π may be ascribed to the celebrated mathematician, Aryabhata I.

118. **Fibonacci** is credited to have introduced the continued fractions to the world in his famous book **Liber Abaci**. The modern practice is, however, to write the continued fraction in descending order as shown here—

Fibonacci

$$x = a_0 + \cfrac{1}{a_1 + \cfrac{1}{a_2 + \cfrac{1}{a_3 + \cfrac{1}{a_4 + \ddots}}}}$$

While giving due credit to Fibonacci, most authors agree that the theory of continued fraction begins with **Rafael Bombelli**, the last of the great algebraist of Renaissance Italy. In his **L'Algebra Opera** (1571) Bombelli attempted to find the square root by means of continued fraction. We can express “2 as a continued fraction, which goes on forever but which has a simple pattern for its components:

$$\sqrt{11} = 3 + \cfrac{1}{3 + \cfrac{1}{6 + \cfrac{1}{3 + \cfrac{1}{6 + \cfrac{1}{3 + \ddots}}}}}$$

119. **Archimedes** of Syracuse (287-212BC) is regarded as the greatest of Greek mathematicians, and was also an inventor of many mechanical devices including the screw, the pulley and the lever. A famous result of his is that the volume of a sphere is two thirds of the volume of its circumscribed cylinder, a picture of which was inscribed on his tomb.

120. **Hypatia** of Alexandria (370-415AD) was the first woman to make a substantial contribution to the development of mathematics. She learned mathematics and philosophy from her father Theon of Alexandria, and assisted him in writing an eleven part commentary on Ptolemy's *Almagest* and a new version of Euclid's *Elements*. About 400 AD, Hypatia became head of Platonist school at Alexandria and lectured there on mathematics and philosophy. Although she had many prominent Christians as students, she ended up being brutally murdered by a fanatical Christian sect that regarded science and mathematics to be worthless to the world.

121. *Carl Fredric Gauss* was one of the most remarkable infant prodigies. His arithmetical powers so overwhelmed his school masters that by the time Gauss was 10 years old, they admitted that

there was nothing more they could teach the boy. It is said that in his first arithmetical class Gauss astonished his teacher by solving what was intended to be a busy work problem – find the sum of all the numbers from 1 to 100? $1 + 2 + 3 + \cdots 100 = ?$

Gauss arranged the numbers making 50 such pairs whose sum is equal to 101

$1 + 100 = 101$

$2 + 99 = 101$

$3 + 98 = 101$

$50+51=101$

Hence the sum of all these pairs $= 50 \times 51 = 5050$.

122. **Euclid,** the father of Geometry has given these five postulates in his famous book Elements.

Postulate 1:- A straight line may be drawn from any one point to any other point.

Postulate 2:- A terminated line can be produced indefinitely.

Postulate 3:- A circle can be drawn with any centre and any radius.

Postulate 4:- All right angles are equal to one another.

Postulate 5:- If a straight line falling on two straight lines makes the interior angle on the same side of it taken together less than two right angles, then the two straight lines, if produced indefinitely, meet on that side on which the sum of angles is less than two right angles.

You would be surprised to know that Euclid did not require his fifth postulate to prove his first 28 theorems. Many mathematicians including him, were convinced that the fifth postulate is actually a theorem that can be proved using the four postulate and other axioms. However, all attempts to prove his fifth postulate as a theorem have failed. But these efforts have led to the creation of Non-Euclid Geometry.

123. **Claudius Ptolemy** in his Almagest had used the symbol for degrees, minutes and seconds but they were different from the modern ones. According to F. Cajori, "The first modern appearance of the degree symbol ^{0} is found in the revised 1569 edition of Gemma Frisius's **Arithmeticae practicae moethodus Facilis**" In **Prvtenicae tabulae coelestium motuum** published in 1571, Erasmus Reinhold used the modern symbol $^{0,\imath,1/2}$ for degree, minute and second.

124. Despite the fact that the authors of *Sulvasutras* were familiar with the right angle triangle, it was Aryabhatta who give the formulae for the area of triangle.

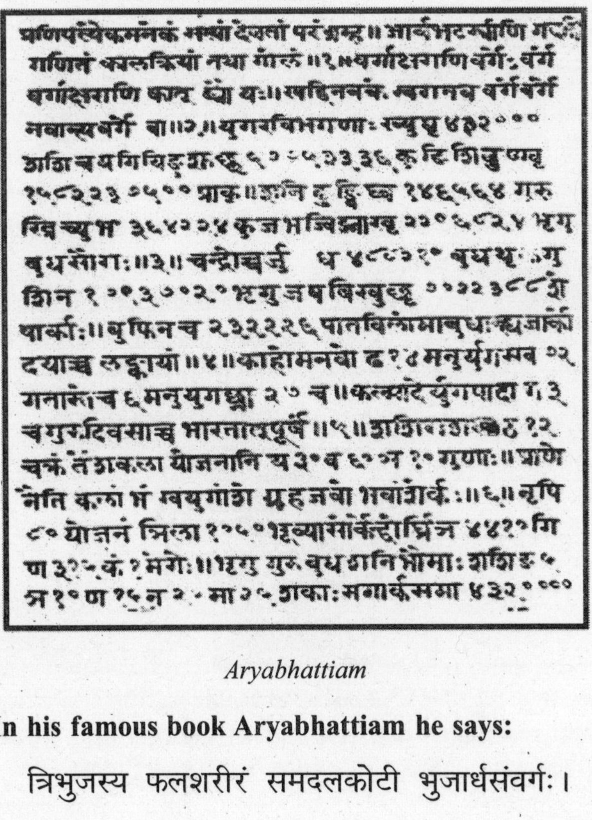

Aryabhattiam

In his famous book Aryabhattiam he says:

त्रिभुजस्य फलशरीरं समदलकोटी भुजार्धसंवर्गः ।

The product of the perpendicular (from the vertex to base) and half the base gives the measures of the area of the triangle.

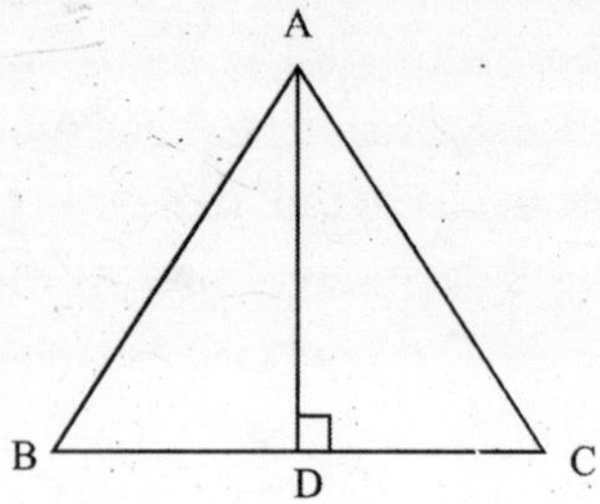

$$ar\,(ABC) = \frac{1}{2} \times BC \times AD.$$

125. **Raman Parimala** an Indian born female mathematician working in University in Atlanta received the 2005 prize in mathematics from the Academy of Science for the Developing world for her work on the quadratic analogue of Serre's conjecture, the Triviality of the Principal homogeneous spaces of classical groups over fields of cohomological dimension 2 and the u invariant of p-adic function fields.

She was born on Nov. 21, 1948 in Tamilnadu, India. She did her M.Sc. from Madras University and Ph.D from Bombay University. She was a Professor at TIFR, Bombay. Since 2005. She is working as a Professor of maths in Georgia.

Raman Parimala

□

3

Symbols

126. $^{2}/_{3} + ^{1}/_{2} - [\ 2+2-3\ \div \{1/2 \text{ of } 4+1-2-(1/4-1/8)\}]$

In the above problem we have four types of brackets

i) small brackets or parenthesis ( )

ii) curly bracket or braces { }

iii) capital bracket [ ]

iv) vinculum —

The first instances of round parenthesis are found in the work of **Nicolo Taratgalia's** (1506-1557) **General trattato di numeri e misure** in 1556.

The first use of braces { } was in **Francois Vieta's Zetetica** in 1593.

In the manuscript edition of **Algebra** by **Rafael Bombelli** (1526-1573) the first use of capital brackets were found. As far as the use of vinculum in mathematics is concerned it was due to **Frans van Schooten** (1615-1660) who while editing **Vieta's** works replaced parenthesis by horizontal bar above.

127. **Indian** algebraists, possibly for the first time in history used abbreviations of names of colours or gems, as symbols of unknown quantities and operations like powers, roots etc.

Present days Notation Times	**Ancient**
Variables x,y,z,...	ma (manika), ni (indranila) mu (mukta)

Operations	
+	yu (yuta)
-	-| (from the Brahmi symbols for k in Ksaya)
×	gu (guna)
/	bha (bhaga)
$(\ )^2$	va (varga)
$(\ )^3$	gha (ghana)
$(\ )^4$	va va
√	mu (mula)

128. Ancient Indian system of measurement of angles was identical to the present system. They used Vikala, Kala and Bhaga for modern second, minute and degree. According to this

60 Vikala = 1 Kala
60 Kala = 1 Bhaga
30 Bhaga = 1 Rashi
12 Rashi = 1 Bhagna

विकलानां कलाषष्टयाः तत् षष्टया भाग
उच्यते तत्रिंशतां भवेद्राशिः भगणो द्वादशैव ते।

(सूर्य सिद्धांत 128)

129. The oldest symbols for the power of a quantity (known or unknown) are found in *UTTARADHYAYA SUTRA*. In it the second power is called Varga (square), the third power is called Ghana (cube), the fourth power is called (varga-varga), and the sixth power is called Ghana-varga, and the twelfth power is called Ghana-varga-varga.

130. In the early times, Indian mathematician used dot to distinguish negative numbers. If a number had no dot it was considered as positive number. In the manuscript of Jainism mathematics –2 was written as 02

131. Our present knowledge of Babylonian achievement in mathematics is due to the extraordinary discoveries of Prof. Otto Neugebauer. Babylonian astronomical texts of the third century BC make explicit use of the following rules:

$$(+) \times (+) = +$$
$$(-) \times (+) = -$$
$$(+) \times (-) = -$$
$$(-) \times (-) = +$$

132 **Diophantus** of Alexandria was the first to make systematic use of symbols in algebraic work, a sign for unknown, a sign for minus, and signs for the various powers, i.e., square, cubes and so on.

133. The Hindu symbol 0 for zero used in inscription and manuscript to indicate the blank position was Sunya, meaning void or empty. This symbol resemble with the Greek symbol the omicron, the first letter of the Greek word ouden ("nothing") . The modern symbol might have been taken from the Greek name omicron used by Buteo in 1559.

134. a:b represents **a** ratio **b** or we can say **a** upon **b** in fractional term where **a** is called antecedent and **b** is called consequent. Though it seems that the symbol of ratio came from ÷ by dropping the bar but there is no historical proof as such to rely upon the fact.

According to DE Smith, the word ratio is a Latin word which was commonly used in the arithmetic of the Middle Ages to mean computation. Smith further says: **To represent the idea which we express by the symbol a:b, the Medieval Latin writers generally used the word proportion, not the word ratio.**

It seems that **Oughtred** used the dot to indicate either division or ratio in his book **Clavis Mathematicae** written in 1631. He used colon (:) to indicate ratio in his book *Canones Sinuum*. In a letter to Bernoulli on July 29, 1698 Leibniz writes : "... in designating ratio I use not one point but two points, which I use at the same time, for division; thus for your dy.x :: dt.a I write dy:x = dt:a; for dy is to x as dt is to a, is indeed the same as, dy divided by x is equal to dt divided by a. From this equation follow then all the rules of proportion."

135. The word congruent is derived from the Latin word Congruere meaning to coincide.

Christopher Clavius in Euclidis volume I writes:-

...Hinc enim fit, ut aequalitas angulorum ejusdem generis requirat eandem inclinationem linearum, ita ut lineae unius conveniant omnino lineis alterius, si unus alteri superponatur. Ea enim aequalia sunt, quae sibi mutuo congruunt.

Leibniz in an unpublished manuscript gives ($\simeq$) to denote the congruence symbol. In **Characteristica Geometrica** Leibniz says that if a figure can be applied exactly to another without distortion, they are said to be congruent.

In Elements of Geometry and Trigonometry a translation of Legendre David Brewster writes:—

Two triangles are equal, when an angle and the two sides which contain it, in the one, are respectively equal to an angle and the two sides which contain it, in the other.

F. Cajori writes that the modern congruent symbol ($\cong$) is due to **Carl Brandan Mollweide** who used it is in **Euclid's Elemente** in 1824. For the congruence of triangles the congruence theorem SSS. SWS and WSW were invented by Julius Worpitzky (1835-1895) where W stands for Winkel =angle.

"When the sides of one plat forme, beareth like proportion together as the sides of any other flatte forme of the same kinde doeth, then are those formes called like flattes .. and their numbers, that declare their quantities, in like sorte are named like flattes.

Robert Recorde in 1557 in Whetstone of Witte writes the above line where **like** word was used in place of similarity. The similarity symbol (~) is due to Leibniz. In his manuscript **Characteristica Geometrica (1679)** [original manuscript has not survived] he wrote **similitudinem ita notabimus: a ~ b.**

136. The perpendicular symbol ($\perp$) was first used by **Pierre Hérigone (1580-1643)** in 1634 in **Cursus mathematicus.** As far as the first use of the word perpendicular in English is concerned, it was due to Chaucer in his treatise on Astrolabe. In geometrical context, it was first used by Sir Henery Billingsley's translation of Euclid's Elements.

137. **Pierre Herigone** (1580-1643) in Cursus $\angle$ mathematicus published in 1634 used and $\angle$ to represent angle. The second symbol $\angle$ had been in use to represent less than was also used for angle.

$\widehat{ABC}$ This symbol was in use during 19th century in Europe.

138. If $x^2 + 4 = 0$, the solution of x cannot be a real number. The fact that the square root of a negative number does not exist in the real number system was recognised by the Greeks. But the Indian mathematician Mahavira was the first to write that the square root of negative number is almost impossible. Leonard Euler was the first to introduce the symbol i (iota) for $i = \sqrt{-1}$

139. The word QED commonly used at the end of proof of the theorem in Geometry stands for **quod erat demonstrandum** (that which was to be proven). Euclid used to conclude his proof with the word **haper edei deiksai** which when translated in Latin by the medieval geometers changed to **quod erat demonstrandem (QED).** The first known of QED in print appears in 1505 in the translation of Euclid by **Bartholemew Zamberti**.

140. Our (+) symbol is a shorthand form of Latin word *et*. It was found in a manuscript dated 1417. Prof D E Smith in

History of Mathematics Volume II page 398 explains

"In a manuscript of 1456 written in Germany, the word *et is* used for addition and is generally written so that it closely resembles the symbol +. The symbol et is also found in many other manuscript as in 5 et 7 for 5+7. "

The minus sign – may have come from the bar – written above m (minus) meaning less. Plus is also called addition and the word addition appears in English in about 1300.

Here tells (th)at (th)er ben .7. spices or partes of (th)is craft. The first is called addicion, (th)e secunde is called subtraccion. The thyrd is called duplacion. The 4. is called dimydicion. The 5. is called multiplicacion. The 6. is called diuision. The 7. is called extraccion of (th)e rote'

The citation is from The Crfte of Nombrynge, one of the earliest manuscripts in the English language that refers to mathematics.

The plus and minus symbols first appeared in print in **Johannes Widmann's Mercentile Arithmetic** where the author says '**was – ist/das ist minus...vnd das + das ist mer**'. But the symbol + and – were not for addition and subtraction but to show the excess or deficiency. Wildmann further speaks:—

'4 Centner + 5 Pfund,' and also of '4 centner –17 pfund' thus showing the excess or deficiency in the weight of boxes or bales.

This book was published in 1489 in Leipzig. F. Cajori in A History of Mathematics writes

'There is a clear evidence that a lecturer at the University of Leipzig Widmann had studied manuscript in Dresden Library in which + and – signify operations, some of these having been written as early as 1486.'

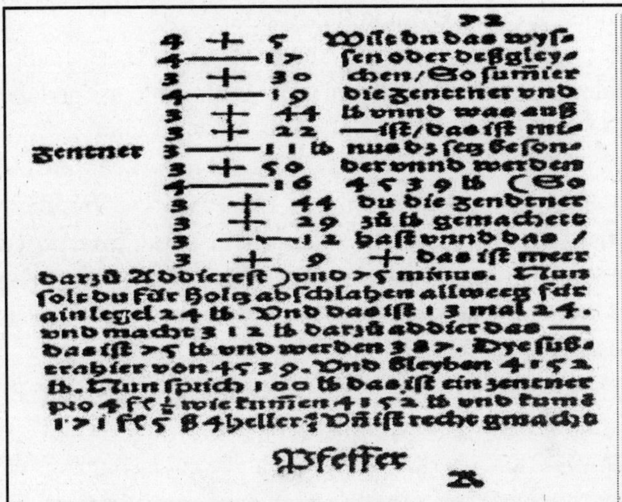

FIRST USE OF THE SYMBOLS + AND —, 1489

The above picture is of **Widmann's Mercentile Arithmetic** published in 1489 in Leipzig. The first person who used + and— in writing an algebraic expression was the Dutch mathematician Giel Vander Hoccke in 1514 whereas the symbol + and – may have been in use in arithmetic to indicate operations by George Walckl. The plus and minus symbols were adopted in the15th century in mathematics after they were used by Robert Recorde in 1557 where he wrote **'There be other two signes in often use of which the first is made thus + and betokeneth more, the other is thus made – and betokenth lesse'.**

141. In the second edition of Edward Wright's translation of Napier's *Descripto* there is a statement—

"The note of addition is + of subtracting (–) of multiplication (X)"

However, the larger symbol (X) was used by William Oughtered in the *Clavis Mathematicae* (Key of Mathematics) written around 1628, but it was not a new mathematical symbol. It was in long use in cross multiplication.

F. Cajori calls (X) as St. Andrew's cross. On account of its resemblance to x. A new symbol (·) **dot** came into use and the first writer who introduced dot in place of X was Leibniz.

In a letter to John Bernoulli, Leibniz writes. **'I don't like X as a symbol for multiplication, as it is easily confounded with x,.. often I simply relate two quantities by an interposed dot and indicate multiplication by ZC.LM. Hence, in designating ratio I use not one point but two points, which I use at the same time for division.'**

Thomas Harriot in a posthumous work **Analyticae Praxis ad Aequationes Algebraicas Resolvwndas** used the dot symbol for multiplication where he wrote 2.aaa =2a cubed but Eves writes: **'Although Harriot on occasion used dot for multiplication, this symbol was not prominently used until Leibniz adopted it.'**

As far as use of word multiplication in English is concerned, it was used in non-mathematical term in 1384 by Chaucer and the present symbol used in mathematics is due to Oughtered.

Here are the two symbols for multiplication; the first on the left hand was used by **Claris** whereas the symbol at the right appears in an anonymous appendix to Edward Wright's translation of Descripto published in 1618. However it is believed to be written by Oughtered.

142. The symbol $\div$

for division was first used by Johann Heinrech Rahn in 1659 in *Teutsche Algebra* which appeared in Zurich and was known in England by Dr. Pell who translated Rahn's work.

In the Ahmes papyrus, the division of a number whose

numerator being 1 is shown in the following way.

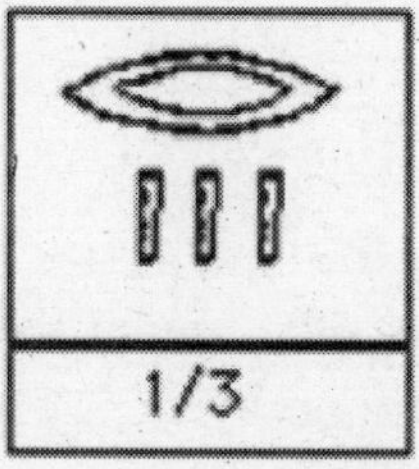

The other symbol for division as 8)24 to mean 24 is divided by 8 was used by **Michael Stifel** in ***Arithmetica* Integra** which was published in 1544. The division symbol 36)116(3; separated by the parenthesis and divisor, dividend and quotient written in the same line was first used by Joseph Ray in 1866 in **Primary Elements of Algebra for Common Schools and Academies.**

143. Per cent had been in use for computing interest, profit and loss problem and tax problem during the end of the 15th century, but they were using the fraction which could be easily reduced in the hundredth. The 15th century Italian manuscript contained the symbol 20 p 100 and x p cento to indicate 20 per cent and 10 per cent. It is believed that the present % symbol has its origin from an anonymous Italian manuscript of 1425 where $\text{P}\ \overline{\ }^{\text{o}}$ is used for percentage.

In the later phase the word **per** was dropped and $\frac{o}{o}$ was used, which finally become %.

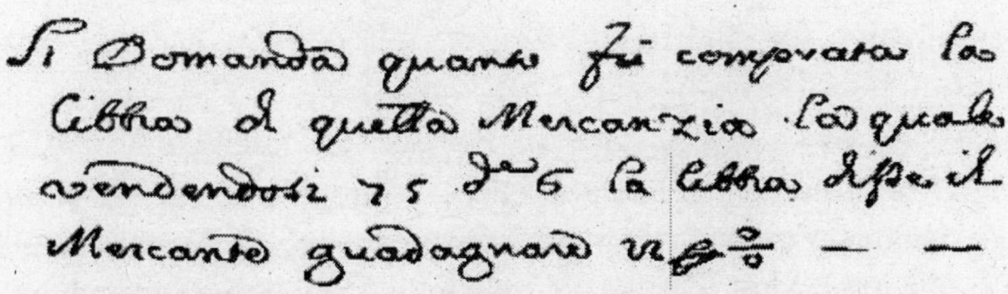

The first apperance of percentage symbol in writing

144. The first use of square root symbol as ℞ was due to **Leonardo of Pisa**. He used this symbol in 1220 in his book **Practica Geometriae**. This symbol was also in use by the late Latin writers.

Chuquet used ℞ 2 and ℞ for square root, ℞ 3 for cuberoot, ℞ 4 for fourth root, and so on.

The radical symbol first appeared in print in 1525 in **Die Coss** by **Christoff Rudolf** (1499-1545), but without the vinculum". It is said that due to its resemblance to the first letter in the word radix the symbol was adopted in mathematics for square root but there is no evidence to prove it.

In **La Geometrie, Rene Descartes** used √ adding a vinculum for square root.

145. One of the most fascinating constant in mathematical history is undoubtedly π, which is used to express the ratio of circumference to diameter. The first Indian mathematician to give the value of pi correct to 4 decimal places was Aryabhatta during the 6th century and since then this constant has become a fascination for mathematics lovers. William Oughtered (1574-1660), Issac Barrow (1630-1677), David Gregory (1661-1701) all used the same notation for pi to designate the circumference of a circle. William Oughtered in *Clavis Mathematicae* used δ/π to designate the ratio of circumference to diameter. J.Christoph Sturm used **e** in place of **π** in his *Mathesis enucleate* published in 1689. William Jones was the first person to use π in his **Synopsis**

palmariorum mathesios in the year 1706 to represent the ratio of circumference to diameter, perhaps, because it is the first letter in the **Perimetron. Euler** adopted this symbol in 1737 and since then it is in use.

146. As stated in History Of Mathematics by D.E. Smith Vol. II page 413, first edition

"The symbol for Infinity ∞ is first found in print in the **Arithmetic Infinitorium** published by Wallis in 1655 and may have been suggested by the fact that the Romans commonly used this symbols for a thousand, just as we use "myriad " for any large number; although in the Greek it meant ten thousand."

Cum enim primus terminus in serie Primanorum sit 0, primus terminus in serie reciproca erit vel infinitus: (sicut, in divisione, si diviso sit 0, quotiens erit infinitus)

The above passage from Arithmetic Infinitorium of John Wallis (1616-1703) has the first use of Infinity which probably he adopted from the late Roman symbol " for thousand.

John Wallis

As no change takes place in the infinite and immutable God when worlds are created or destroyed, though numerous orders of beings are absorbed or put forth, so as when a quantity is divided by zero the result is an infinite quantity.

The above lines are from **Bhaskaracharya's Lilavati** clearly advocated the term infinity meant for a/0 = ∞.

147. According to Todhunter's translation of Euclid Book I definition 15 says:—

"A circle is a plane figure bounded by one line which is called circumference."

The circle with a dot in the centre is due to Heron around 150 AD whereas the circle without dot was due to Pappus.

In the Hindu and Buddhist mythology the circle $\odot$ represent the female power and the dot at the centre represent the male, and the circle with the dot at the center represent the merging of the two forces together.

According to Smith, **Boethius** seems to have been the first to use the word semi diameter. The Radius was used by Ramus word (1515-1572). As far as the **R** and **r** are concerned, **Euler** used these two symbols to indicate the radius of the circumscribed and inscribed circle respectively.

148. **"I will sette as I doe often in woorke use, a paire of parralles, or Gemowe lines of one lengthe, thus : ==, bicause noe 2, thynges, can be moare equalle."**

The above passage from **Whetstone of Witte** was written by **Robert Recorde** (1510-1558) in 1557 shows the first use of equality sign in mathematics as it is used today.

This symbol did not gain immediate popularity and two vertical lines were widely accepted; and during 1700, the words **ae** or **oe** taken from the word **acqualis** meaning equal were highly used.

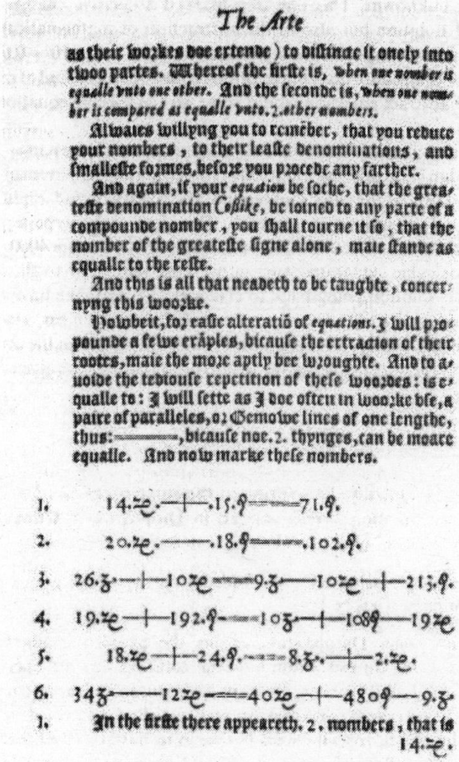

The Arte

as their workes doe extende) to distincte it onely into twoo partes. Whereof the firste is, *when one nomber is equalle unto one other.* And the seconde is, *when one nomber is compared as equalle unto .2. other nombers.*

Alwaies willyng you to remēber, that you reduce your nombers, to their leaste denominations, and smalleste formes, before you procede any farther.

And again, if your *equation* be soche, that the greateste denomination *Cosike*, be ioined to any parte of a compounde nomber, you shall tourne it so, that the nomber of the greateste signe alone, maie stande as equalle to the reste.

And this is all that neadeth to be taughte, concernyng this woorke.

Howbeit, for easie alteratiō of *equations.* I will propounde a fewe exāples, bicause the extraction of their rootes, maie the more aptly bee wroughte. And to auoide the tediouse repetition of these woordes: is equalle to: I will sette as I doe often in woorke use, a paire of paralleles, or Gemowe lines of one lengthe, thus: =======, bicause noe .2. thynges, can be moare equalle. And now marke these nombers.

1. 14.ℨ. + .15.ϑ ======= 71.ϑ.
2. 20.ℨ. — .18.ϑ ======= .102.ϑ.
3. 26.ʒ. + 10ℨ ======= 9.ʒ — 10ℨ + 213.ϑ.
4. 19.ℨ + 192.ϑ ======= 10ʒ + 108ϑ — 19ℨ
5. 18.ℨ + 24.ϑ. ======= 8.ʒ. + 2.ℨ.
6. 34ʒ — 12ℨ ======= 40ℨ + 480ϑ — 9.ʒ.

1. In the firste there appeareth. 2. nombers, that is 14.ℨ.

Whetstone of Witte

This symbol of equality can be seen in **Ahmes Papyrus**, the first book available till know written around 3500 BC. The equal symbol did not appear in print until 1618. It appeared in an anonymous Appendix in the English translation of **Napier's Descripto of Edward Wright**.

149. While comparing between two numbers such as 21>19 or 19 < 21; we generally use > (greater than) or < (less than) symbols. The use of these symbols is not confined to arithmetic alone because while comparing two sides in any geometrical figure we apply these symbols.

We all are aware of the two postulates:

i) The sum of two sides of a Δ is greater than the third side. If we express this in mathematical symbol we can say $a+b>c$ where a, b and c are the sides of a Δ.

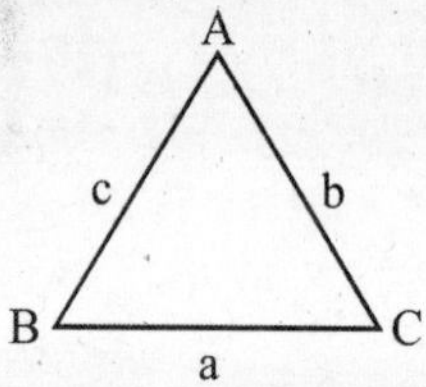

ii) The angle opposite to greater side is greater.

Clearly these postulates use the concept of greater than or less than or vice versa.

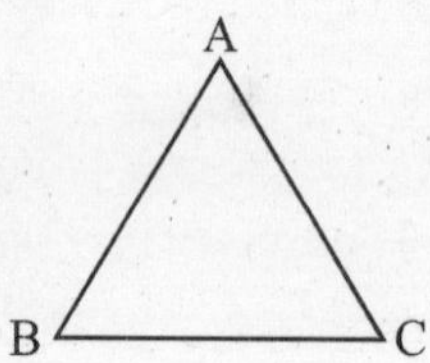

If $\langle B \rangle \angle C \Rightarrow AC > AB$

Thomas Harriot, an English mathematician (1560-1621), while surveying the North American city in 1585 saw a Native American with this symbol  on his arm and probably developed the above two symbols from this one.

Signum majoritatis ut $a > b$ significet a majorem quam b" and "Signum minoritatis ut $a < b$ significet a minorem quam b.

In **Artis Analyticae Praxis ad Aequationes Algebraicas Resolvendas** (The Analytical Arts Applied to Solving Algebraic Equations) written by **Thomas Harriot the** symbols $<$ and $>$ first appears, which was published posthumously in 1631.

 This symbol meant for **A major est quam B**

 And this symbol meant for **A minor est quam B**

can be seen in a book ***Lectiones Opticae & Geometricae*** written by Issac Barrow (1630-1677).

As far as the symbols and for less than or equal to and greater than or equal to are concerned, **Pierre Bougeur** (1698-1758), a French geodesist invented these symbols in 1734 but it is not known who invented the present symbols e" for greater than or equal to and d" for less than or equal to but it is apparent that the modern symbols are the refinement of the older symbols used by **Bougeur**.

150. In the **Ahmes Papyrus** there are several examples on fraction but those are of unit fraction.

Suppose if 5 loaves were to be divided among 12 persons then the Egyptians would have given one third and one twelfth to each man.

$^5/_{12} = ^1/_3 + ^1/_{12}$

The Babylonians made extensive use of sexagesimal fractions.

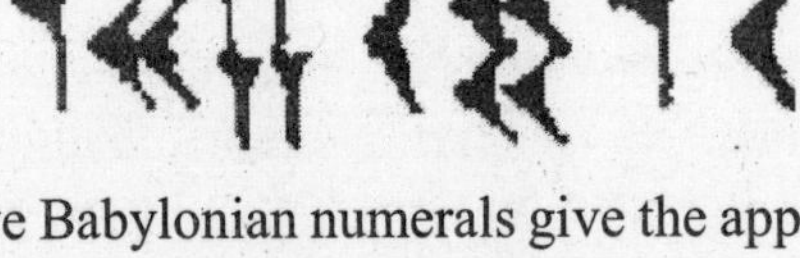

The above Babylonian numerals give the approximate value of $\sqrt{2}$ =1.414222. The Egyptians had the symbols for unit fractions. They wrote the symbol  over the number to represent a fraction. Thus  =1/3 and  would represent 1/10. The only other fraction for which they had a symbol was 2/3. The Romans employed **duodecimal** fractions since their weight and money were divided in terms of 12 parts. The Romans used words to indicate fraction. A unit of weight was the **as and the unica** was a **twelfth**

part of as. According to **DE Smith,** Indian mathematicians **Brahmagupta** and **Bhaskara** wrote fraction as we use today but without the bar. The horizontal fractional bar is due to **al-Hassar** around 1200. F. Cajori writes that **Fibonacci** was the first European to use fractional bar as it is used today.

151. The word decimal is derived from the Latin word Decimus meaning Tenth. In the early printed books numbers in multiple of 10 were called decimal numbers. The word decimal appears in English I 1608 in the title Disme: The Art of Tenths or Decimall Arithmetike a work of Simon Stevin translated by Robert Norman. DE Smith writes 'Franceso Pellos in the second volume of his commercial arithmetic published in 1492 at Turin used decimal point to denote the division of a number by a power of ten.'

Christoff Rudolf in 1530 used the vertical bar to denote decimal. Smith further writes:

The first man who gave evidence of having fully comprehended the significance of all this preliminary work seems to have been Christoff Rudolf, whose Exempel Büchlin appeared at Augsburg in 1530. In this work he solved an example in compound interest, and used the bar precisely as we should use a decimal point today. If any particular individual were to be named as having the best reason to be called the inventor of decimal fractions, Rudolff would seem to be the man, because he apparently knew how to operate with these forms as well as merely to write them, as various predecessors had done. His work, however, was not appreciated, and apparently was not understood, and it was not until 1585 that a book upon the subject appeared.

Though Napier used comma to separate the whole part from the fractional part, but decimal point first appears in the 1771 edition of Encyclopedia Britannica in the article Arithmetick.

152. "When measuring angles, mathematicians don't use

degrees, they use radians," "There are 2 pi radians in a circle. This means one quarter of a circle corresponds to half of pi. Tau, the 19th letter of the Greek alphabet **Tau** (uppercase **T**, lowercase **τ**), was chosen independently as the symbol for 2 pi by Michael Hartl, physicist and mathematician and author of "The Tau Manifesto," and Peter Harremoës, a Danish information theorist, Bob Palais of the University of Utah. In "Pi is Wrong!" Palais argued that, for thousands of years, humans have been focusing their attention and adulation on the wrong mathematical constant.Two times pi, not pi itself, is the truly sacred number of the circle, Palais contended. The mathematician are therefore looking for a new symbol Tau that will in future replace the old symbol pi = 3.14 and place the new constant Tau =6.28, the true ratio of circumference to diameter. May be, in coming years, you will witness the new symbol, defining the ratio of circurfence to diometer.

□

4

Origin of words

153. The Hindu name for zero used in inscriptions and manuscripts to indicate the blank position was **Snunya,** meaning void or empty. The Arabic text used the word Sifr meaning vacant. Abraham ben Meir ibn Ezra (1092-1167) used galgal for zero in a description on a decimal system of numeration. In the Latin translation of the Arabic text it was called **zephirum. Dionysius Exiguus** used the word nulla to indicate zero in his Easter tables. In Fibonacci's ***Liber Abaci*** zephirum for zero was used whereas Italians used zeuro. Cipher is also found in English as early as 1399. The first printed treatise on zero *De arithmetrica opusculum* was written in 1491 by Philippus Calandrus. Maximus Planudes called it tziphra and this was used by Fine (1530) in 16th century.

In the further journey zero was given the name such as zepiro, cifra which finally led the foundation of the word zero. The modern symbol of zero by 0 might have been taken from the Greek name omicron used by **Buteo** in 1559.

9TH century	11th century	12th century Pali manuscript	12th to 16th century Jainism manuscript	INTERNATIONAL	DEVNAGRI	TIBETAN	MANGOLI	SINGHLI	BERMISE	SYAMI	JAVI	KHUMER	ARABIC
0	O	[illegible]	[illegible]	0	0	∘	O	[illegible]	O	O	∘	**0**	·

154. The word **mathematics** is derived from a Greek word **μάθήμα** (Mathema) meaning a subject of instruction; but by the time of Aristotle the term was used definitely restricted to subjects of the quadrivium. Pythagoras is said to have coined the words philosophy for "love of wisdom" and mathematics for "that which is learned."

155. **ALGEBRA** comes from the title of a work written in Arabic about 825 by al – Khwarizmi al-jabr w'al-muqâbalah. Al-jabr means the reunion of broken parts. When this was translated from Arabic into Latin four centuries later, the title emerged as Ludus algebrae et *almucgrabalaeque.* The earliest known use of the word algebra in English in its mathematical sense is by Robert Recorde in *The Pathwaie to Knowledge in 1551.*"

156. **ARITHMETIC** is a Greek word transliterated into English as arithmetike. It passed into Latin as arith-metica. According to Smith (vol. 2, page 8):

The word arithmetic, like most other words, has undergone many vicissitudes. In the Middle Ages, through a mistaken idea of its etymology, it took an extra **r,** as if it had to do with "metric." So we find Plato of Tivoli, in his translation (1116) of Abraham Savasorda, speaking of Boetius in arismetricis. The title of the work of Johannes Hispalensis, a few years later (c. 1140), is given as "Arismetrica," and fifty years later than this we find Fibonacci dropping the initial and using the form "Rismetirca." The extra r is generally found in the Italian literature until the time of printing. From Italy it passed over to Germany, where it is not uncommonly found in the books of the 16th century, and to France, where it is found less frequently. The ordinary variations in spelling have less significance, merely illustrating, as in the case with many other mathematical terms, the vagaries of pronunciation in the uncritical periods of the world's literatures.

157. The term **TRIGONOMETRY** is due to Bartholomeo Pitiscus (1561-1613) and was first printed in his Trigonometria: sive de solutione triangulorum tractatus brevis et perspicuus, which was published as the final part of Abraham Scultetus Sphaericorum libri tres methodicé conscripti et utilibus scholiis expositi (Heidelberg, 1595)

The word first appears in English in 1614 in the English translation of the same work: Trigonometry: or The Doctrine of Triangles. First written in Latin, by B. Pitiscus, and now translated into English, by Ra. Handson.

158. In ancient India, conventional mathematics termed **Ganitam** was known before the development of algebra. This is borne out by the name–Bijaganitam, which was given to the algebraic form of computation. Bijaganitam means the other mathematics (Bija means another or second and Ganitam means mathematics). The fact that this name was chosen for this system of computation implies that it was recognised as a parallel system of computation, different from the conventional one which was used since the past and was till then the only one. Some have interpreted the term Bija to mean seed, symbolising origin or beginning. And the inference that Bijaganitam was the original form of computation is derived. Aryabhatta, in his treatise Aryabhattiya has mentioned about Bijaganitam. Bhaskaracharya has also authored a treatise on this subject. The treatise which is dated around the 12th century A.D. is entitled *Siddhanta-Shiromani* of which one section is titled Bijaganitam.

159. The word **mile** comes from the Roman milia, thousands. The Romans measured distances in paces, which were about five feet. So, milia passum, 1,000 paces or about 5,000 feet, was the length of a mile. The length of a mile used to be 5,000 feet. However, in 1575 the British Parliament added 280 feet to this measurement, declaring the mile to be 1,760 yards or 5,280 feet, so

that it could be divided evenly into furlongs. One furlong is 660 feet long, giving 8 furlongs to the mile.

160. **Algorithm,** as it is used in mathematics means a systematic procedure to solve a problem. The word is derived from the name of the Persian mathematician, al-Khwarazmi.

161. **Arithmetic** was the Greek word for number, and is closely related to the root of **reckon**, which is becoming an obsolete term for count. In the middle ages the best mathematicians of Germany were called **Reiche-nmeister** and their arithmetic texts were Reichenbucher The beginning of the word is drawn from the Indo-European root ar which is related to "fitting together. The first arithmetic book published in North America was *Sumario compendioso de las quentas de plata y oro que in los reynos del Piru son necessarias a los mercaderes y todo genero de tratantes Los algunas reglas tocantes al Arithmetica.* The title translates to Comprehensive Summary of the counting of silver and gold, which, in the kingdoms of Peru, are necessary for merchants and all kinds of traders. The author was Brother Juan Diez.

162. The Latin root of the word **circle** is circus. The traditional shape of the large roofless enclosures in which the famous Roman chariot races were run was circular or oblong, and thus the word came to describe this shape as well. The word **centre** comes to us from a Greek root kentrus, for a spur or sharp pointed object.

163. **Fraction** comes from the Latin word frangere, to break. A fraction, then, originally represented the broken portion of some whole. The first known use of the word in English is by Geoffrey Chaucer in 1391 in the work, *A Treatise on the Astrolabe.*

164. The word logarithm is the combination of two words logos and arithmos. The meaning of these Greek words are ratio

and number respectively. Hence logarithm means ratio number.

165. The word Geometry comes from Geo and Metre meaning "Measuring the Earth".

166. Probability theory had its origin in 16th century when an Italian mathematician Jerme Cardon wrote his first book on the subject ***Book on Games of Chance (Liber de Ludo A leae)*** which got published in 1663 after his death.

167. The origin of the word **Calculus** is from the Greek word Kalyx used for pebble or small stone. According to Encyclopedia Britannia, “About 15 BC, the Roman architect and engineer Vitruvius mounted a large wheel of known circumference in a small frame, in much the same fashion as the wheel is mounted on a *wheelbarrow;* when it was pushed along the ground by hand it automatically dropped a pebble into a container at each revolution, giving a measure of the distance traveled. It was, in effect, the first odometer.”

These odometers were used in taxi carriages. Each time the wheel of the carriage turned, a pebble, a calculus, dropped from a container into another. In the end of the ride, the driver counted how many pebbles had dropped, and that determined the price of the transportation. This kind of usages of pebbles gave the word *Calculus* its present meaning. The word Volume meaning capacity used in mensuration comes from the Latin word Volvere meaning to turn or roll. The idea comes from the practice of writing on scrolls that were then rolled into a cylinder. According to Steven Schwartzman's The Words of Mathematics, the use of the word for an amount of space began around the 16th-17th century.

□

5
Books

168. The oldest book in mathematics, *Ahmes Papyrus* now better known as *Rhind Papyrus* was named after the Scotsman Henery Rhind, who found it in an Egyptian shop in 1858.

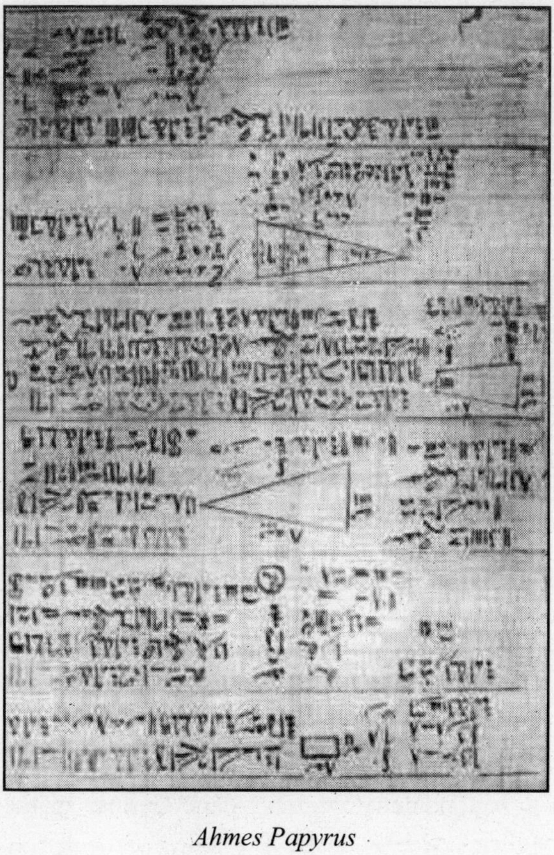

Ahmes Papyrus

Ahmes Papyrus or the Rhind Papyrus, the main source of information of Egyptian mathematics which is now kept in the British Museum was written about 2200 BC. This papyrus is called the Ahmes Papyrus in the honour of Ahmes who was responsible for copying it from the ancient mathematical book. It has 80 problems on Algebra, Geometry, Mensuration, etc.

The other oldest book in Mathematics known as Moscow Papyrus has its origin in 1700. BC. and contains 25 Problems. It is now kept in **Moscow Museum of Fine Art.**

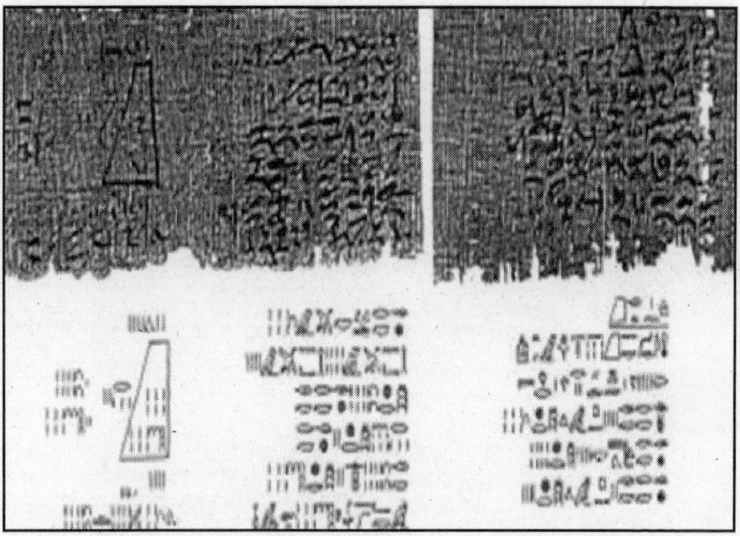

Moscow Papyrus

169. In **Surya Siddhanta,** Bhaskaracharya calculated the time taken for earth to orbit the sun to 9 decimal places 365.258756484 days which is just differed by 0.00085 days of modern accepted measurement of 365.2596 days.

170. *De triangulues Onimodis Libri* was the first work solely devoted to trigonometry written by the German mathematician Johann Muller.

171. **Euclid,** the father of geometry, wrote 13 volumes of Elements, at the age of 40. *Elements*, the most respectable book in the history of mathematics contains the following topics

- I Congruence, Parallel lines, Pythagoras theorem
- II Geometric treatment of identity $(a+b)^2 = a^2 + b^2 + 2ab$, areas, the golden section
- III Circles
- IV Inscribed and Circumscribed polygon
- V Proportion treated geometrically
- VI Similarity of polygon
- VII –IX Arithmetic
- X Incommensurable magnitude
- XI-XIII Solid geometry

172. The advanced Chinese mathematics book Kiu-chang-suan-shu or Arithmetic in nine sections deals with the following topics:—

Fangtien (–squiring the farm) relating to areas
Sumi (calculating the cerels) relating to % and proportion
Shuai-jen (calculating the shares) partnership and rule of three
Shao-kuang—relating to find the length of square
Shang-kung—finding volumes
Kin-shu—problem of allegation
Ying-pu-tsu—addition and subtraction
Fang-cheng—relating to simultaneous equation and some concept of determinant
Kou-ku—relating to right angle triangle

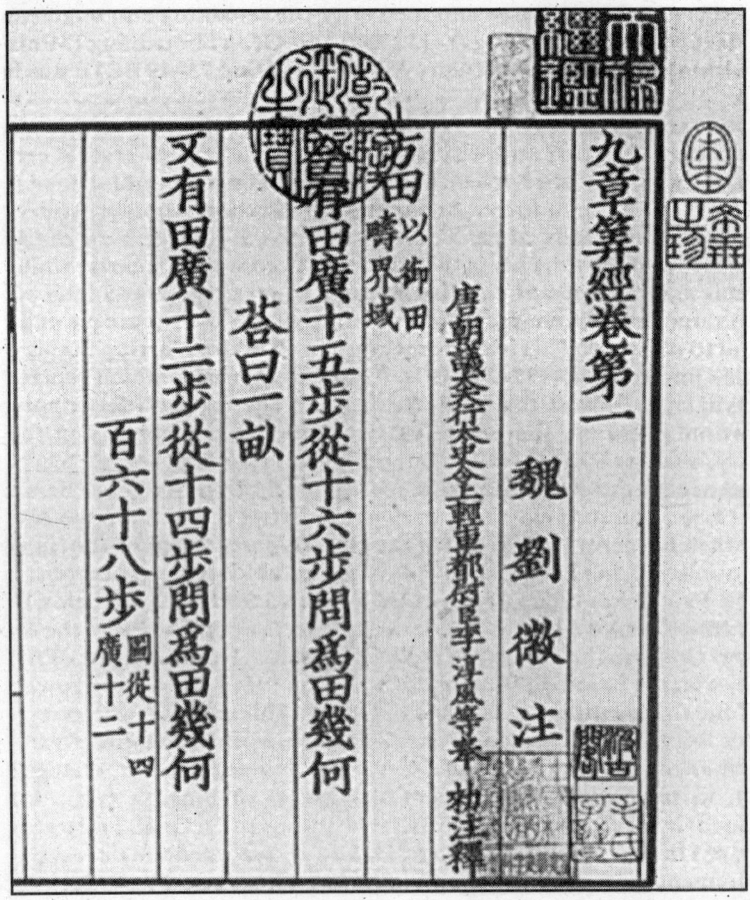

九章筭經卷第一

魏 劉 徽 注

唐朝議大夫行太史令上輕車都尉臣李淳風等奉 勑注釋

方田 以御田疇界域

今有田廣十五步從十六步問爲田幾何

荅曰一畝

又有田廣十二步從十四步問爲田幾何

百六十八步 圖從十四 廣十二

Chapter 1 of the Nine Chapters on the Mathematical Art

This practical book contains 246 problems on different topics in mathematics.

173. ***Almagest,*** the great work of Ptolemy has thirteen volumes. The original Greek title of the book is ***Tne Greatest Compilation.*** When the book was translated into Arabic it got its name ***al-majisti*** and when again translated into Latin it got its name as ***Almagest***.

174. In a book published in 1940, there are 370 different proofs of the famous Pythagorean Theorem; among them is one proved by the American President James Garfield.

James Garfield

175. The first treatise on spherical trigonometry. *The book of unknown arcs of a sphere* was written by al-Jayani. It contains formulas for right handed triangles △ Solution of Spherical. △ by means of polar triangles.

176. The oldest extant mathematical manuscript in South Asia is the *Bakhshali Manuscript*, a birch bark manuscript written in Sarada script. This manuscript was discovered in 1881 by a farmer while digging a stone enclosure in the village Bakhshali, near Peshawar (now in Pakistan). The surviving manuscript has seventy leaves, some of which are in fragments. The topics treated in this book include **Fractions, Square Roots, Profit and Loss, Simple Interest, the Rule of Three, Simultaneous linear equations, Quadratic equations and Arithmetic Progression.**

The manuscript is now preserved in the Bodlein Library at Oxford University.

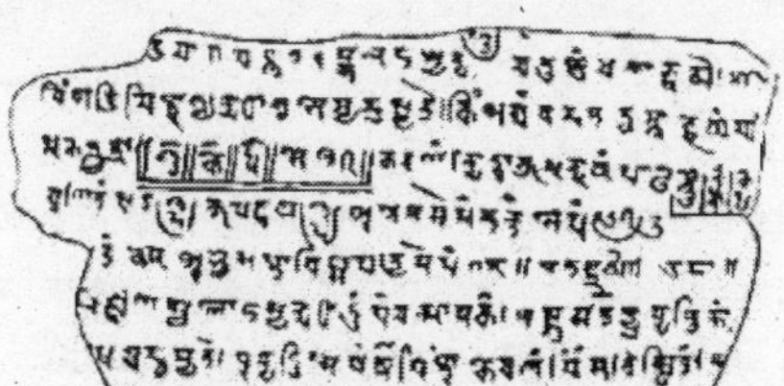

□

Bakshali manuscript

6
Puzzles

177. The population of Rabbits grows in accordance with Fibonacci series

1, 1, 2, 3, 5, 8, 13…

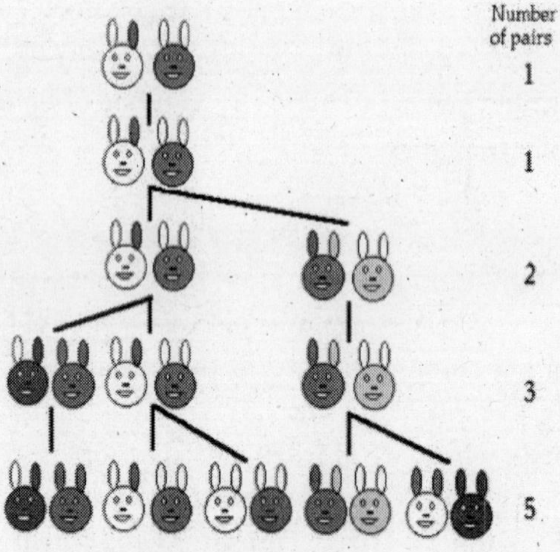

In Fibonacci series every next number of the series is the sum of the previous two.

178.

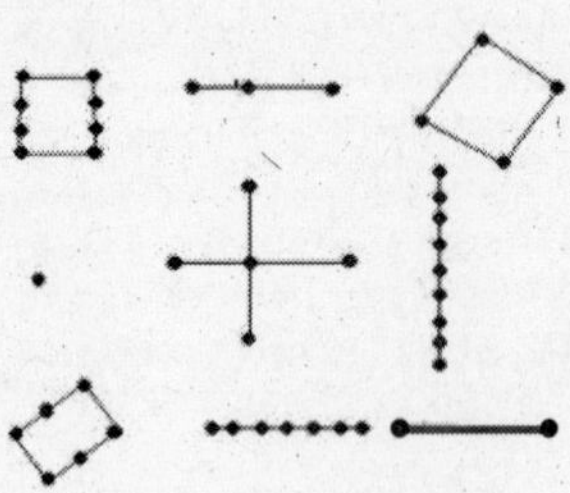

This is a Losu Symbol. It was found on the back of a tortoise in China in 12th century. This is the oldest magic square whose every row, every column and both diagonals totals at 15.

8	3	4
1	5	9
6	7	2

There is a very interesting story related to the magic square. Once the Chinese King Yu was walking along the Lo River (Yellow river), he saw a mystical turtle crawling on the river bank. The turtle was ordinary in all respect except that its shell had a series of dots within a square. This amused the king Yu as the sum of each row, column and diagonal was 15. Yu took the turtle to his palace where the mystical turtle spent the rest of its life.

179. The male bee has 1 parent the female bee has 2 parents, 3 grand-parents, 5 great-grand-parents, 8 great-great-grand-parents, and 13 great-great-great-grand-parents. Here the numbers 1, 1, 2, 3, 5, 8, 13 forms a Fibonacci sequence. All the females are produced when the queen has mated with a male and so have two parents. So female bees have 2 parents, a male and a female whereas male bees have just one parent, a female.

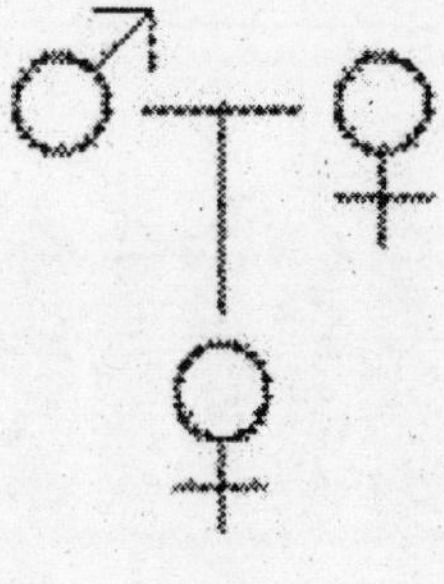

Queens have 2 parents

Males have 1 parent

180. 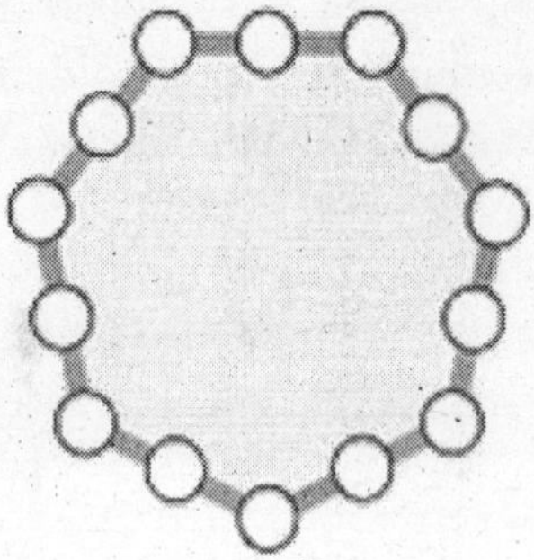

The Heptagonal puzzle was invented by Henry Dudeney (1847-1930). In order to solve it one needs to place the numbers from 1 to 14 in every circle so that the three numbers along every side add up to 26.

181. Suan pan is considered as the earliest abacus with beads on rods. The term Suan –pan means calculating plate. It has two beads above a reckoning bar and five beads below the reckoning bar. The first device which was used is mathematics for calculation was **abacus**. Though its origin is unknown but probably Chinese are credited for its development.

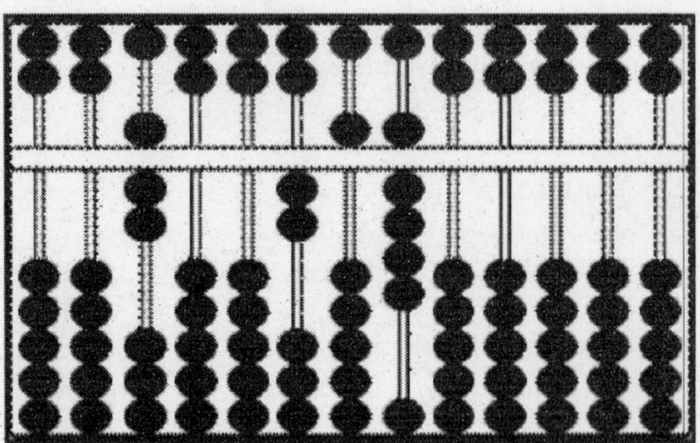

Suan Pan

182.

333667× 223 = 741741741
33336667×22233= 741174117411
3333366667×222333=741117411174111
333333666667×2223333=741111741111741111
33333336666667×22233333=741111174111117411111

183. Take any three digit number with its entire digit different. Now reverse the number and subtract it from original number taken. Once again let the number so obtained be reversed and add it to the number obtained in second step. What did you get? It is 1089. Let us see the pattern of 1089 and its reversal:—

1089 ×9 = 9801
10989 × 9 = 98901
109989 × 9 = 989901
1099989×9 = 9899901
10999989 × 9 = 98999901
109999989 × 9 =989999901
....
....

184. The invention of the tangram puzzle is unrecorded in history. The earliest known Chinese book is dated 1813 but the puzzle was very old by then. One reason for this could be that in China, its country of origin, at the time it was considered a game for women and children. Tangrams enjoyed a surge of interest during the 19th century in Europe and America. It interests the maths-inclined with the geometry and ratios of the pieces.

185. The famous 15-Puzzle was invented by **Sam Lloyd** (1841-1911). He has often been referred to as America's greatest puzzlist.

1	2	3	4
5	6	7	8
9	10	11	12
13	14	15	

186. **Harry Nelson** was the first person to produce the 3×3 magic squares containing only consecutive primes for which he won a $100 prize offered by Martin Gardner.

1480028159	1480028153	1480028201
1480028213	1480028171	1480028129
1480028141	1480028189	1480028183

187.

987654321	1
87654321	12
7654321	123
654321	1234
54321	12345
4321	123456
321	1234567
+ 21	12345678
1	+ 123456789
1083676269	1083676269

188. The square of the middle of any three consecutive numbers is always 1 more than the product of the first and the last number.

Suppose the three consecutive numbers chosen by you are:—

11 12 13

Square of the middle = $12^2 = 144$

Product of the 1st and 3rd = $11 \times 13 = 143$

189. The oldest mathematical puzzle is found in one of the oldest mathematical books called **Ahmes Papyrus**.

As I was going to St. Ives, I met a man with seven wives, each wife had seven sacks, each sack had seven cats, and each cat had seven kits. Kits, cats, sacks and wives, how many were going to St.Ives.

This sum requires the knowledge of geometric progression.

Man	1
Wife	7
Sack	49
Cat	343
Kit	2401
Total	**2801**

190. There are 26 alphabets in English and it can make 40329×10^{22} different combinations.

A	**B**	**C**	**D**	**E**	**F**
G	**H**	**I**	**J**	**K**	**L**
M	**N**	**O**	**P**	**Q**	**R**
S	**T**	**U**	**V**	**W**	**X**
Y	**Z**				

191. Look at the following magic square:—

8811	8188	1111	1888
1118	1881	8818	8181
8888	8111	1188	1811
1181	1818	8881	8118

The square total is 19998 in all directions. This is better known as IXOHOXI magic square, because when you see the image of this magic square in the mirror, you will find the new magic square inside the mirror.

8811	8188	1111	1888
1118	1881	8818	8181
8888	8111	1188	1811
1181	1818	8881	8118

8881	1111	8818	1188
1818	8188	1881	8111
1181	8811	1118	8888
8118	1888	8181	1811

The most important thing about the mirror magic square is that its magic constant also comes out to be 19998.

192.

9×9+7=88

98×9+6=888

987×9+5=8888

9876×9+4=88888

98765×9+3=888888

987654×9+2=8888888

9876543×9+1=88888888

98765432×9+0=888888888

193.

12345679 ×9 = 111111111

12345679×18 = 222222222

$12345679 \times 27 = 333333333$

$12345679 \times 36 = 444444444$

$12345679 \times 45 = 555555555$

$12345679 \times 54 = 666666666$

$12345679 \times 63 = 777777777$

$12345679 \times 72 = 888888888$

194.

31

331

3331

33331

333331

3333331

33333331

Look at the above magical patterns of prime numbers. The next prime number in this series is 17 times 3's with a 1 at the end. The above series is particularly attractive because the numbers of threes in the first seven primes increase by one. There are only eleven other primes in this series with less than 1800 series.

195.

$0 \times 9 + 1 = 1$

$1 \times 9 + 2 = 11$

$12 \times 9 + 3 = 111$

$123 \times 9 + 4 = 1111$

$1234 \times 9 + 5 = 11111$

$12345 \times 9 + 6 = 111111$

$123456 \times 9 + 7 = 1111111$

$1234567 \times 9 + 8 = 11111111$

$12345678 \times 9 + 9 = 111111111$

196.

$11 \times 11 = 121$

$111 \times 111 = 12321$

$1111 \times 1111 = 1234321$

$11111 \times 11111 = 123454321$

$111111 \times 111111 = 12345654321$

$1111111 \times 1111111 = 1234567654321$

$11111111 \times 11111111 = 123456787654321$

$111111111 \times 111111111 = 12345678987654321$

197.

$7 \times 7 = 49$

$67 \times 67 = 4489$

$667 \times 667 = 444889$

$6667 \times 6667 = 44448889$

$66667 \times 66667 = 4444488889$

$666667 \times 666667 = 444444888889$

$6666667 \times 6666667 = 44444448888889$

198.

$987654321 \times 9 - 1 = 8888888888$

$87654321 \times 9 - 1 = 788888888$

$7654321 \times 9 - 1 = 68888888$

$654321 \times 9 - 1 = 5888888$

$54321 \times 9 - 1 = 488888$

$4321 \times 9 - 1 = 38888$
$321 \times 9 - 1 = 2888$
$21 \times 9 - 1 = 188$
$1 \times 9 - 1 = 08$

199. This is a Nasik Magic square.

1	14	7	12
15	4	9	6
10	5	16	3
8	11	2	13

This is also known as perfect or diabolic square. The magical constant is 34. The best part of this magic square is that if you repeat the square in all directions, you can then draw a box around 4×4 array of numbers and it will be magic. The 4×4 array numbers that has been highlighted also represents a magic square and to your astonishment, the magic constant in this case is also 34. A Nasik magic square remains Nasik magic square under four different transformations:- rotation, reflection, a transfer of a row from top to bottom or from bottom to top or a transfer of a column from one side to another. Isn't it a beauty!

1	14	7	12	1	14	7	12
15	4	9	6	15	4	9	6
10	5	16	3	10	5	16	3
8	11	2	13	8	11	2	13
1	14	7	12	1	14	7	12
15	4	9	6	15	4	9	6
10	5	16	3	10	5	16	3
8	11	2	13	8	11	2	13

200. Give a piece of paper to your friend and tell him to do the following calculation.

- Ask him to write any 7 digit number of his choice.
- Multiply it by 8.
- Add 7 digit number + 8 + Number × 8

Sum up the digits individually until you get a single digit number.

The final answer will always be 8.

Example: Let your number = 3104804

Multiply it by 8 = 3104804 × 8 = 24838432

Now add seven digit number + 8 + Previous result = 3104804 + 8 + 24838432 = 27943244

Sum up the digit individually = 2 + 7 + 9 + 4 + 3 + 2 + 4 + 4 = 35 = 3 + 5 = 8

Properties of 8:—

i. It is the largest cube in the Fibonacci sequence.
ii. One can split a pie into 8 pieces with three straight cuts.
iii. It is a cube number 23 = 8
iv. 5 + 1 + 2 = 8 and 512 = 83
v. A cube or cuboid has 8 vertices.
vi. An octopus has eight arms.
vii. If the last three digits of a number is 000 or is divisible by 8 the number is divisible by 8.
viii. An umbrella is octagonal in shape.
ix. Many spiders, bees and flies have eight legs.

Spider

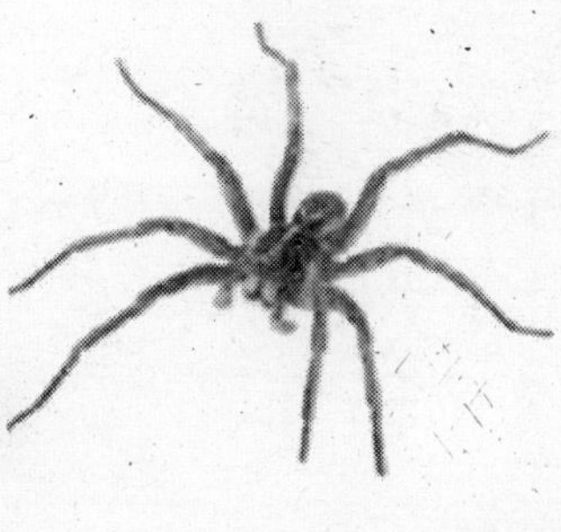

Bee

Iichneuman fly

Crane fly

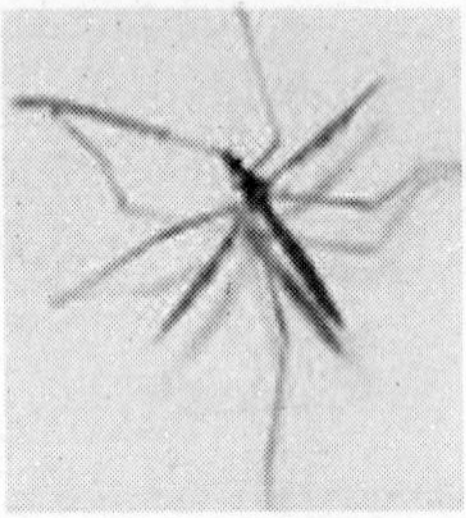

□

7

Theorems and Curves

201. Pascal's triangle was first introduced by the Chinese mathematician Yang Hui but 500 years later it was again re-discovered by Blaise Pascal.

```
                  1
               1     1
            1     2     1
         1     3     3     1
      1     4     6     4     1
   1     5    10    10     5     1
1     6    15    20    15     6     1
...   ...   ...   ...   ...   ...   ...
```

Pascal Triangle

202. The Pythagoras theorem states– In a right angles triangle, the square of hypotenuse is equal to the sum of the squares of other two sides.

Pythagoras

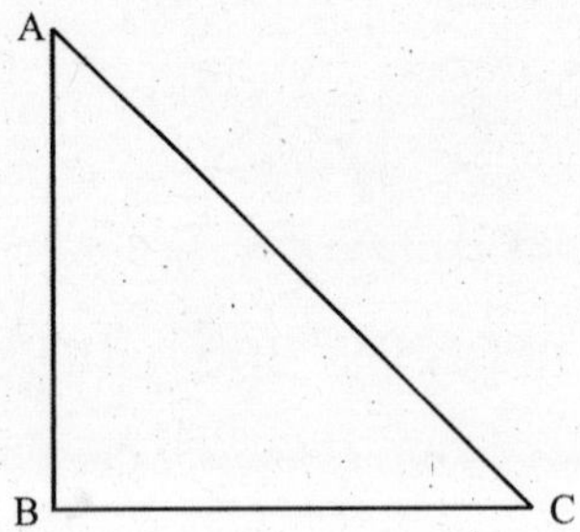

In the above right angle triangle,

$AC^2 = AB^2 + BC^2$

The Pythagoras theorem $x^2+y^2=z^2$ was well known to Indian much before the Pythagoras proved it. The Sulva-Sutra of Apastamba gives rules for constructing right-angled triangles of following lengths (3,4,5), (5,12,13), (8,15,17) ,(12,35,37), (12,16,20) and (15,20,25). The Arabs used the term Figure of Bride and the Greeks used the Theorem of married women for Pythagoras theorem perhaps because it represents to unite two in one.

Pythagoras is credited to have discovered the irrational number $\sqrt{2}$. **Theodorns of Cyrene** who is said to have been Plato's teacher had discovered the irrational $\sqrt{3}$, $\sqrt{5}$, $\sqrt{7}$, $\sqrt{8}$, $\sqrt{10}$, $\sqrt{11}$, $\sqrt{12}$, $\sqrt{13}$, $\sqrt{14}$, $\sqrt{15}$ and $\sqrt{17}$.

203. Conic Section (Parabola, Hyperbola, and Ellipse) was discovered by Menaechmus.

Parabda *Ellipse* *Hyperbola*

Equation of Parabola = $y^2 = 4ax$

Equation of Ellipse = $\frac{x^2}{a^2} + \frac{y^2}{b^2} = 1$

Equation of Hyperbola = $\frac{x^2}{a^2} - \frac{y^2}{b^2} = 1$

204. The Chinese text, *Chou Pei* presents the oldest known proof of the right-angle triangle theory in the hsuan-thu diagram. This theory is commonly known as the Pythagorean theorem,

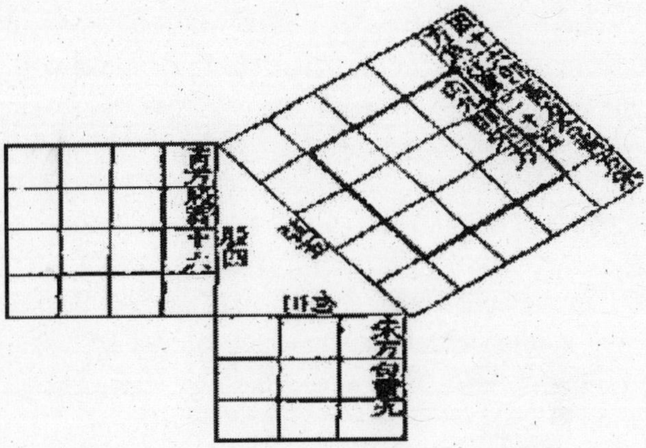

Here, $3^2 + 4^2 = 5^2$

Pythagoras Theorem in Chinese Book

205. The first printed book on Arithmetic was published in Italy in 1478, whereas the first modern book on the theory of numbers **Arith Medicae** was published in 1801 by Carl Friedrich Gauss.

206. The credit to expand the binomial theorem $(a+x)^n = {}^nc_0a^x + {}^nc_1a^{n-1}x^1 + {}^nc_2a^{n-2}x^2 + \ldots\ldots\ldots + {}^nc_nx^n$ where n, a positive number goes to Newton.

Issac Newton

He stated the Binomial Theorem without proof. He published *Philosophiae Naturalis Principia Mathe-matica* (1687). This is one of the most important contributions to the field of science. This work was revised by Roger Cotes in 1713 and later translated into English by Andrew Motte. In 1704, Newton published *Opticks* in which he revealed unique discoveries on the properties and characteristics of light.

207. In the mid-1990s, Texas banker Andrew Beal posed a perplexing mathematical problem and offered \$5,000 for the solution of this problem. The value of the prize increases by \$5,000 per year up to \$50,000 until it is solved. In particular, Beal was curious about the equation $A^x + B^y = C^z$. The six letters represent integers, with x, y, and z greater than 2. (Fermats Last Theorem involves the special case in which the exponents x, y, and z are the same.) Oddly enough, Beal noticed that when a solution of this general equation existed, then A, B, and C have a common factor. For example, in the equation $3^6 + 18^3 = 3^8$, the numbers 3, 18, and 3 all have the factor 3. Using computers at his bank, Beal checked equations with exponents up to 100 but could not discover a solution that did not involve a common factor.

208. Three famous construction problems of antiquity are:—

(i) Squaring the circle

(ii) Duplication of cube

(iii) Trisecting the angle using only a straightedge and compass.

Although mathematicians were unable to solve the problem of trisecting the angle with straightedge and compass alone, they were able to trisect it by other means. Many of these solutions used curves other than circles. For example Hippias (425BC), a great scholar, trisected the angle using a curve called quadratrix. Archimedes (250BC) achieved a trisection of the angle using a spiral; other mathematicians used other curves, including the conic section. Archimedes also demonstrated a way to trisect the angle using only a compass and a straightedge with two marks on it. Many mathematical discoveries were made as a result of attempts to trisect the angle. Yet despite the best efforts of the best mathematicians, the problem remained unsolved. Finally, in the19th century it was realised that it would always remain so. The problem was algebraically proved impossible by Wantzel (1836).

209. The Greek **Eratosthenes** devised a laborious method to find out prime numbers between two numbers, known as "Sieve of Eratosthenes". Refinements of this method have been used to compute the primes up to about 10,000,000.

Sieve of Eratosthenes

Eratosthenes (276-195BC) gave a golden rule to trace the primes between two given numbers. Though this method is very simple, it cannot be regarded as useful in practice as it is very lengthy and time-consuming.

First write down the number from 2 to N. Remove all the multiples of 2 and 3. Continue the process until all the multiples of primes not greater than square root of N have been removed.

Suppose we have to find the primes below 30. First find the square root of 30 and that is 5.477... (approximately), so try to remove all the multiples up to prime 5.

②	③	~~4~~	⑤	~~6~~	7	~~8~~	~~9~~
~~10~~	11	~~12~~	13	~~14~~	~~15~~	~~16~~	17
~~18~~	19	~~20~~	~~21~~	~~22~~	23	~~24~~	~~25~~
~~26~~	~~27~~	~~28~~	29	~~30~~			

Hence the primes below 30 are 2, 3, 5, 7, 11, 13, 17, 19, 23 and 29.

Eratosthenes

Eratosthenes (276 BC-194 BC) is considered to be the first to calculate the Earth's circumference. He did so using arithmetic calculations and information on the varying altitudes of the sun at noon over Alexandria and Syene (now Aswan, Egypt). Eratosthenes one of the greatest scholars of Alexandria who was called **'The Second Plato'** by his admirer had his nickname **Beta** which signified him the second of the wise men of antiquity .

210.

The Pascal triangle developed by French mathematician Blaise Pascal is not only useful in finding the binomial expansion but it has many unusual properties and a variety of uses:-

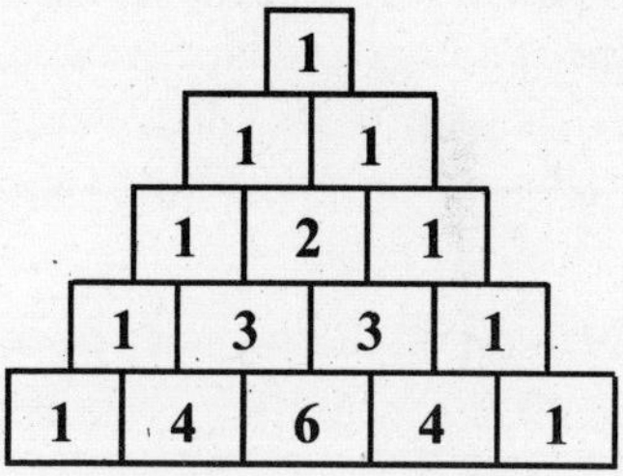

- Horizontal rows add to the powers of 2.
  Sum of 2nd row = 1 + 1 = 2 = 2^1
  Sum of 3rd row = 1 + 2 +1 =4 =2^2
  Sum of 4th row = 1 + 3 + 3 +1 =8 =2^3
  Sum of 5th row = 1 + 4 + 6 + 4 + 1 =16 = 2^4
- The horizontal rows represent power of 11
  First row = 11
  Second row = 121 = 11^2
  Third row = 1331 = 11^3
  Fourth row = 14641 = 11^4
- The numbers on diagonals of the triangle add to the Fibonacci series 1, 1, 2, 3, 5, 8, 13, 21, 34...
- Pingala, a writer in 3rd century BC in his book Chanda-Sastra had Described about Meru-Prastara now called Pascal triangle.

211. The last theorem of Fermat states that

$$x^n + y^n = z^n$$

has no non-zero integer solutions for x, y and z when $n > 2$. The full text of Fermats statement, written in Latin, reads "cubum autem in duos cubos, aut quadrato-quadratum in duos quadrato-quadratos, et generaliter nullam in infinitum ultra quadratum potestatem in duos eiusdem nominis fas est dividere cuius rei demonstrationem mirabilem sane detexi. Hanc marginis exiguitas non caperet" (Nagell 1951, p. 252).

In translation, it is impossible for a cube to be the sum of two cubes, a fourth power to be the sum of two fourth powers, or in general for any number that is a power greater than the second to be the sum of two like powers. I have discovered a truly marvellous demonstration of this proposition that this margin is too narrow to contain.

This theorem remained the mystery for mathematicians for over 350 years when Andrew Willes came with a final proof of this theorem in 1995.

Postage stamp when Fermats Theorem was solved

212 The equation $Nx^2+1 = y^2$ is better known as **Diophantine equation** of second degree. This problem was discussed for the first time in the history of maths, to solve such equation was developed by Bhaskara II.

There is an interesting history behind an equation of the type $Nx^2+1 = y^2$

In 1657, the French Mathematician Fermat proposed to his friend Frenide to solve the intermediate equation

$$61x^2+1=y^2$$

But neither of them succeeded in solving the equation. Euler gave a complete solution but surprisingly the very same equation was completely solved by Bhaskara II in his book *Siddhanta*

Sniromani way back in 1150. Bhaskara obtained the value of x and y respectively in 9 to 10 digits for $61x^2+1=y^2$ and they were x= 226153980 and y=1766319049.

213. This shape below is called the **spiral.** This spiral is called **Archimedes Spiral**. It was first discovered by Archimedes.

Spider's web

The space between each line of the spiral and the one before and after is the same. The polar equation of this spiral is $\mathbf{r = a\theta}$. This spiral was first studied by Archimedes in 225 BC. The inner structure of a sea shell also looks like a spiral.

This spiral is called **equiangular or logarithmic spiral**. The curve of the spiral always intersects the outreaching radii at a fixed angle. Logarithmic spirals also occur in the curve of the elephants' tusk, the horns of the wild sheep and even in the canary's claws. The name equiangular spiral was given by Rene Descartes.

□

8

Other Fields

214. International Mathematical Olympiad was introduced in the year 1959. The first IMO was held in Romania, since then it had been held every year except 1980. It is a two-day maths competition held each summer. Participating countries send teams of up to 6 students.

215. Regular polyhedron e.g. a tetrahedron (4 sides), an octahedron (8 sides), an icosahedron (20 faces), a cube (6 faces) and dodecahedron (12 pentagonal faces) are known as PLATONIC solids in honour of the Great Greek Philosopher Plato who associated these figures with the universe and its four elements fire, water, air, and earth.

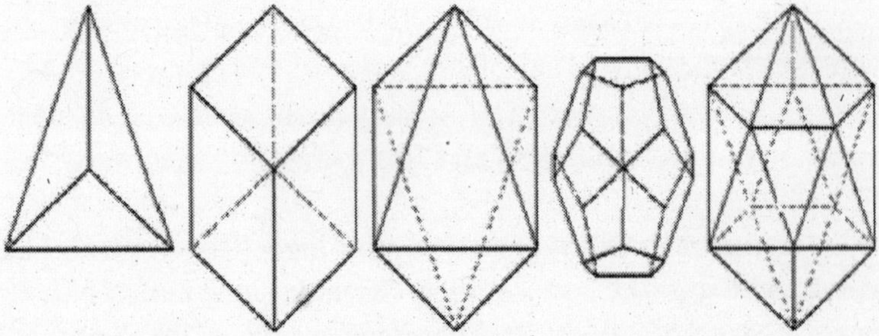

Platonic Solids

216. Among all the shapes with the same perimeter, the circle has the longest area. Among all the shapes with the same area, the circle has the shortest perimeter.

217. One does not need the English alphabet **'a'** while writing the names of +ve integers (one, two, three...) until one reaches one thousand.

218. Every triangle has 9 special points associated with it. These are 3 middle points, three altitudes and three middle points of lines joining vertices to the orthocenter of the triangle, and all these nine points lie on a circle called NINE POINT CIRCLE.

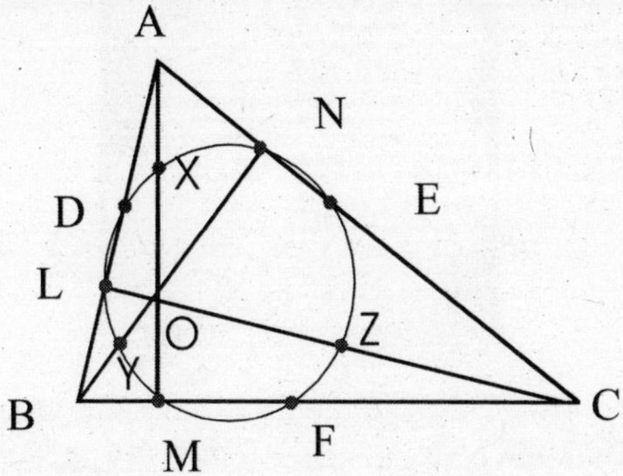

In the above picture D, E and F are the midpoints of side AB, BC and AC. CL, AM and BN are the altitudes drawn from the vertex C, A and B on the opposite sides. O is the orthocentre. Moreover, X, Y and Z are the mid-points of AO, BO and CO. The nine points D, E, F, L, M, N, X, Y, Z lie on a circle. This result was known by Euler in 1765, but re-discovered by Feuerbach in 1822. The resulting circle is known as the Feuerbach circle or Nine Point Circle.

219. The oldest summing counting board is the Salamis tablet used by the Babylonians. It was discovered in 1846 on the island of Salamis. Initially it was thought to be a gamming board. This is made of white marble and now kept at the National Museum of Epigraphy in Athenes.

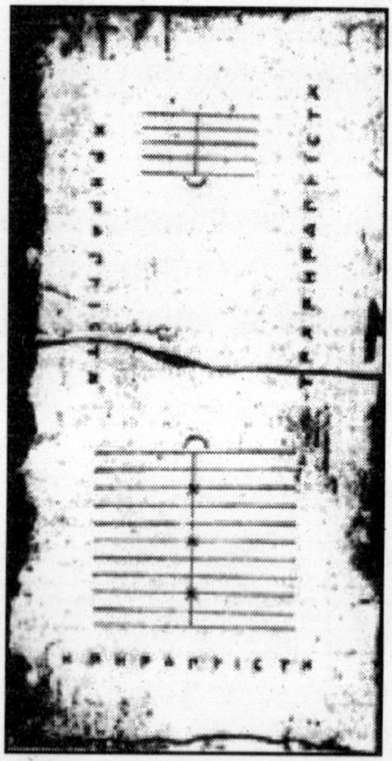

220. In 1862, Yale University of United States of America became the first university to award a PhD degree in mathematics.

221. **Symmetry** can be noticed everywhere in Nature. There are different kinds of symmetry but most common of them is called Bilateral symmetry. If you divide such a symmetrical figure vertically in the middle, then both the halves will be identical.

Butterfly

Leaf

Deer

222. The first work on Newtonian Calculus written by Charles Hayes was published in London in English language in 1704.

223. The earliest history of the introduction of Egyptian calendar of twelve months each of 30 days and 5 feast days is dated back to 4241 BC.

224. The Towers of Hanoi Puzzle was invented in 1883 by Edouard Lucas, a French mathematician. This puzzle was originally sold as a toy, and was described as a simplified version of the Tower of Brahma.

The legend has it that the mythical Tower of Brahma, in the Indian city of Benares, has a post with 64 gold discs stacked in decreasing sizes. These discs are said to be moved, one at a time, by the priests of the Hindu temple from the original post to one of the other two posts, with a larger disc never being placed on top of a smaller disc. It was said that by the time all 64 discs have been transferred to another post in the original order, the temple will crumble to dust and the earth would disappear. At the rate of moving 1 disc per second around the clock, it is estimated to take about 585 billion years to complete the transfer of the gold discs in the Tower of Brahma. This was calculated using the formula: $X = 2^n - 1$, where X is the number of moves required and where n is the number of discs to be moved.

A model of Tower of Hanoi

225. Mathematics-based subject Operational Research had its birth during Second World War. The term 'Operations Research' is known as "Operational Research" in Britain and other parts of Europe. Other terms used are "Management Science," "Industrial Engineering," and "Decision Sciences". The multiplicity of names comes primarily from the different academic departments that have hosted courses in this field. The subject is frequently referred to simply as "OR", and includes both the application of past research results and new research to develop improved solution methods.

Operations Research can be defined as the science of decision-making. It has been successful in providing a systematic and scientific approach to all kinds of government, military, manufacturing, and service operations. Operations Research combines mathematics, statistics, computer science, physics, engineering, economics, and social sciences to solve real-world business problems.

226. The first mathematics society London Mathematical Society was established in 1865.

This Society was established by Augustus De Morgan on 16 Jan 1865, who was also its first president. The Society publishes books and periodical, organizes mathematical conferences, provides funding to promote mathematics research and also awards a number of prizes for excellence in mathematical research.

227. The Fields Medals are commonly regarded as mathematics closest analog to the Nobel Prize (which does not exist in mathematics), and are awarded every four years by the International Mathematical Union to one or more outstanding researchers. 'Fields Medals' are more properly known by their official name, "International medals for outstanding discoveries in mathematics."

The Field Medals were first proposed at the 1924 International Congress of Mathematicians in Toronto, where a resolution was adopted stating that at each subsequent conference, two gold medals should be awarded to recognise outstanding mathematical achievement. Professor J. C. Fields, a Canadian mathematician who was secretary of the 1924 Congress, later donated funds establishing the medals which were named in his honour. Consistent with Fields wish that the awards recognise both existing work and the promise of future achievement, it was agreed to restrict the medals to mathematicians not over forty at the year of the Congress. In 1966, it was agreed that in light of the great expansion of mathematical research, up to four medals could be awarded at each Congress.

The Fields Medal is the highest scientific award for mathematicians, and is presented every four years at the International Congress of Mathematicians, together with a prize of 15000 Canadian dollars. The first Fields Medal was awarded in 1936 at the World Congress in Oslo.

The Fields Medal is made of gold, and shows the head of Archimedes (287-212 BC) together with a quotation attributed to him: Transire suum pectus mundoque potiri (Rise above oneself and grasp the world). The reverse side bears the inscription: Congregati ex toto orbe mathematici ob scripta insignia tribuere. (The mathematicians assembled here from all over the world pay tribute for outstanding work).

Archimedes on Fields Medal

228. The Egyptian mathematicians had the knowledge of advanced mathematics which can be seen with the dimension of pyramids they had built. The maximum error in the length of the sides of the great pyramid is only 0.63 inches which is 1/14000 of the total length and also the error in the angle at the pyramid corner is no more than 12" of a degree i.e. 1/27000 of the angle 90°. Such accuracy cannot be attained without the vast knowledge of advanced mathematics.

Pyramid

229. India has given the world two most important things in mathematics:– a) Zero b) Decimal system.

Decimal system is a way of writing numbers. Any number, from huge quantities to tiny fractions, can be written in the decimal system using only ten symbols (0-9). Each symbol has an absolute value and the value of each symbol is determined by its place value. The Vyasa Bhashya write:—

यथा एकरेखा शतस्थाने शतं दशस्थाने दशैवं चैकस्थाने यथा च एकत्वेपि स्त्री माता च उच्यते दुहिता स्वसा च इति।

In the unit place, the digit has the same value; in the 10th place 10 times the value; and in 100th place 100 times the value as a woman is called mother, daughter and sister.

Laplace writes:—

It is India that gave us the ingenious method of expressing all the numbers by ten symbols, each symbol receiving a value of position, as well as an absolute value. We shall appreciate the grandeur of this achievement when we remember that it escaped the genius of Archimedes and Appolonius.

230. "Hau, its whole, its seventh, it makes Nineteen". This is one of the oldest problem in mathematics given in Ahmes Papyrus. "In modern English it means that there is a number to which when one seventh of it is added the result is nineteen." The word **Hau** denoted an unknown quantity in those days.

231. Gustave Eiffel included the names of 72 French Scientists on plaques around the first stage when he built the famous Eiffel Tower.

Eiffel Tower

Some of the mathematicians whose name have been included are:—

1. Ampere 2. Cauchy 3. Columb 4. Fourier
5. Lagrange 6. Laplace 7. Legendre 8. Poisson

232. Two tablets found in Senkerah in 1854 dated 2000 BC show the amazing aspect of the Babylonian's calculating skills. The square of numbers up to 59 and cubes of number up to 32 are written on these tablets. YBC 7289 or 'root two tablets' of Babylon demonstrated the Babylonian knowledge of square root.

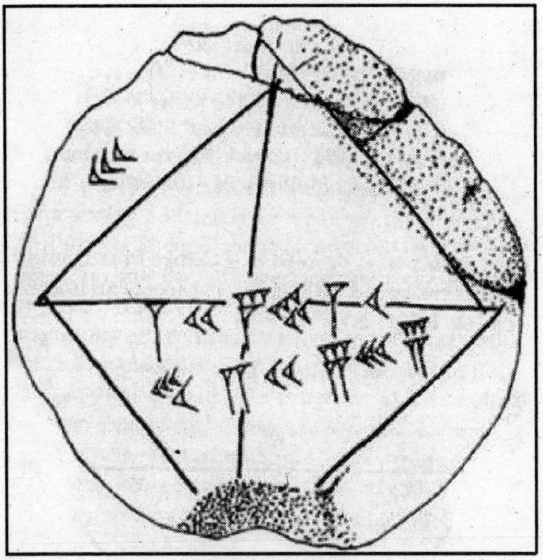

YBC 7289 Tablet

233. By using the four weights 1 Kg, 3 Kg, 9 Kg, and 27 Kg one can weigh any object between 1 to 40 Kg.

334. If you assign A=1, B=2, C=3, ... and Z=26 then P+R+I+M+E = 16+18+9+13+5+=61 is also prime number.

235. Nature also exhibits different geometrical shapes. Apple, when cut through the middle always shows a five point star.

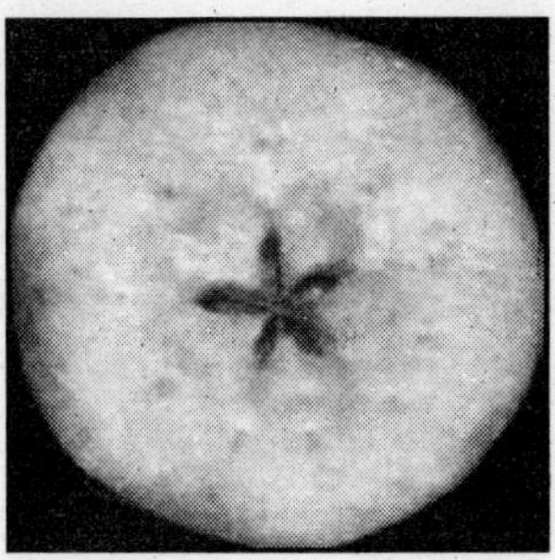

Apple

Oranges look like spheres but when cut through the middle, the shape of triangles is visible.

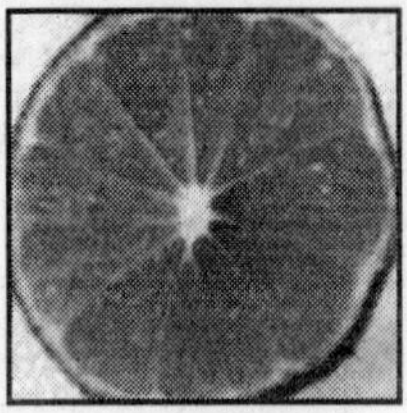

Orang

The cross-section of bamboo is hexagonal in shape.

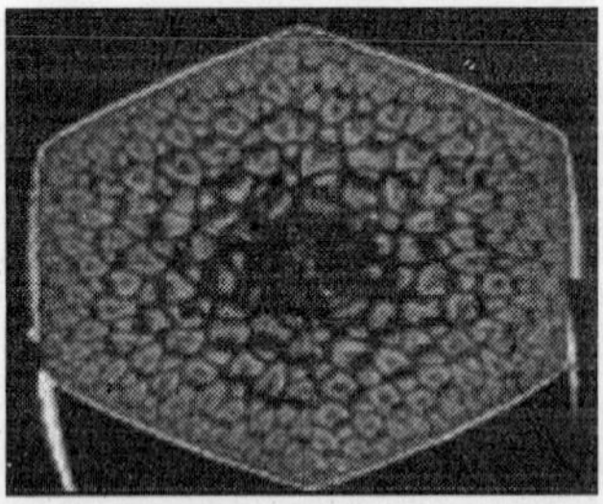

Bamboo

Trees planted at the two sides of road represent a parallel line. The two banks of the rivers also exhibit the example of parallel line.

Besides these, there are many more shapes in Nature and now it is for the readers to observe and find others.

236. This is interesting to know that even animals have shown sense of numbers in the experiments conducted over them. The number sense allows an animal to determine the difference in size between two small collections of similar objects.

Birds have shown that they can be trained to determine the number of seeds in different piles of seeds up to the number five. An example of their ability to count up to five is contained in the story of the crow. In this story, a squire was very frustrated with a particular crow. This crow had made it's nest in the watch tower of the squires estate. The squire continuously tried to surprise the crow, but had no such luck. Each time the squire approached the nest, the crow would leave. It would not return until the squire had left. From a distance the bird would watch for the mans departure. With this in mind, the squire came up with a plan to fool the bird. He took another man into the tower with him. Hoping to outwit the bird, one man left the tower. However, the bird would still not return to the tower until both men had departed. This experiment was repeated in the succeeding days with two, three, and even four men. Yet, the crow was not to be fooled. Finally, five men were sent into the tower. As before, all entered the tower. Only one remained as the other four left the area. At this point, the crow lost count and returned to its nest. Not being able to distinguish above the number four cost the crows life, as the squire rid the tower of the crow forever.

237. In the construction of temple of Athens Parthenes the Pythagorean Theorem was used. The length, breadth and height of the temple were 69.5 m, 30.88 m and 13.72 m respectively. To the astonishment of the mathematician, the ratio comes out to be 16:36:81

238. There are few number in mathematics that exhibit special propenliis:

"The sum of squares of the two halves of the number is equal to the number itself"

$12^2 + 33^2 = 1233$

$990^2 + 100^2 = 990100$

$9412^2 + 2353^2 = 94122353$

$74160^2 + 43776^2 = 7416043776$

$116788^2 + 321168^2 = 116788321168$

$88^2 + 33^2 = 8833$

Explanation:- In the first example 1233 has four digits. You can divide it into two halves namely 12 and 33. The square of these halves when added gives the original number. You can extend it for other examples

(12) (33)

$12^2 + 33^2 = 144 + 1089 = 1233$

239. In a positional numeration system (place-value system) the numerical value of each symbol or digit depends on its position in the row of digits.

Every integer larger than 1 can serve as a base in a place-value system and the system requires as many digit symbols as indicated by the numerical value of the base.

Base	Name of System
2	Binary
3	Ternary
4	Quaternary
5	Quinary
6	Senary
7	Septenary
8	Octanery
9	Nonary
10	Decimal
11	Undenary
12	Duodecimal
16	Hexadecimal
20	Vigesimal
60	Sexagesimal

240. Mr. Rajan Srinivasan Mahadevan continuously cited the value of pi over 31800 places in 3 hrs 49 mins.

241. A figure with 20 sides is called Icosahedron. It is interesting to note that it is impossible to draw a regular solid with more than 20 sides.

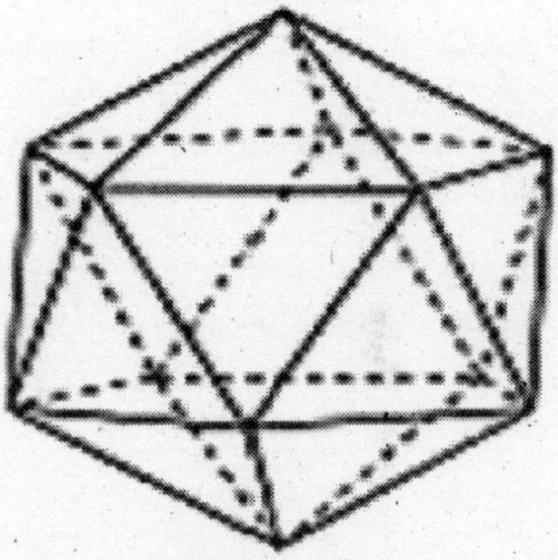

Icosahedron

242. Shakuntala Devi, the human computer as people call her, is a mathematical genius.

Shakuntala Devi

On June 18, 1980, she demonstrated the multiplication of two 13-digit numbers picked at random by the Computer Department of Imperial College London, and answered it in 28 seconds, faster than the Computer.

$$\begin{array}{r} 7{,}686{,}369{,}774{,}870 \\ \times\ \underline{2{,}465{,}099{,}745{,}779} \end{array}$$

18,947,668,177,995,426,462,773,730

Shakuntala Devi, a calculating prodigy was born on Nov 4, 1939.

243.

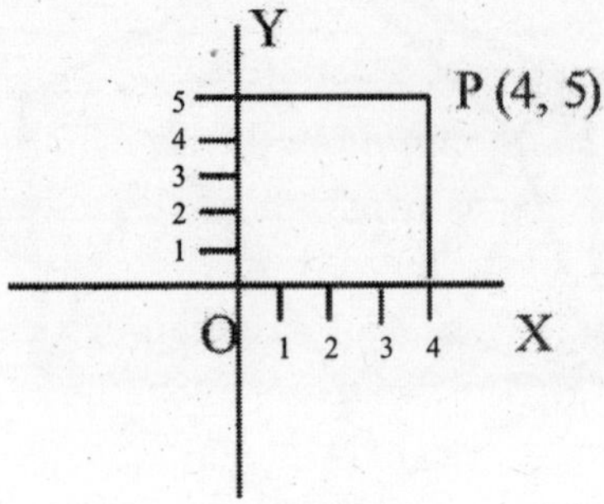

The above graphical representation shows that P (4, 5) lies in the first quadrant. The x-co-ordinate of P is called the abscissa and the y-co-ordinate of P is called the ordinate. The Cartesian graphical representation was first done by Rene Descartes. Modern analytical geometry is called Cartesian Geometry after the name of Rene Descartes (1596-1650). The above term abscissa, ordinate and co-ordinate is due to Leibnitz.

244. Will you believe my words, if I say that the sum of angles of a triangle may not be 180° in all the cases? I know it will be hard to assume. In Euclidean geometry, what we read today, the sum of angles of triangle is always equal to 180°, but in Non-Euclidean Geometry, the sum of angles of a triangle may be more or less than 180°. In Lobachevsky- Bolyai geometry the sum of angles of a triangle is always less than 180° whereas in Riemannian geometry the sum of angles of a triangle is always greater than 180°.

Lobachevsky-Bolyai Triangle

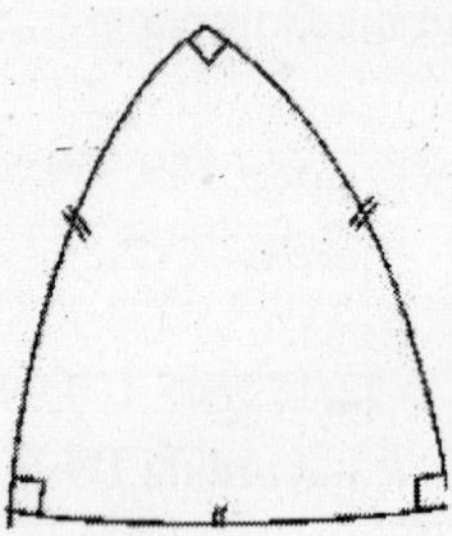

Riemannian triangle

Lobachevsky-Bolyai geometry is based on the assumption that at least two lines can be drawn through a given point parallel to a given point; on the other hand, the Riemannian geometry is based on the assumption that no lines are ever parallel drawn through a fixed point manually at right angles to each other.

245. The modern period of mathematics dates from the beginning of the 19th century and its dominant figure is C.F. Gauss. In the area of geometry, Gauss made fundamental contributions to differential geometry, did much to found what was first called analysis situs and now better known as Topology and anticipated by the breakthrough of non-euclidian Geometry though he could not get his result published. This breakthrough was made by Lobachevsky and Bolyai. Each of them proceeded by establishing the independence of Euclid's fifth postulate and showing that a different, self consistent geometry could be derived by substituting another postulate in its place. Still another non-Euclidean geometry was invented by Bernhard Riemann, whose work also laid the foundation for the modern tensor calculus description of space, so important in the general Theory of Relativity.

246. It is hard to believe that mathematicians are equally intelligent in other fields too. Look at the following table and see how mathematicians are being awarded Nobel Prize for their contributions in other fields:—

Name of Mathematician	Year	Field
Lorentz	1902	Physics
Planck	1918	Physics
Einstein	1921	Physics
Russel	1950	Literature
Tinbergen	1969	Economics
Kantorovich	1975	Economics
Selten	1994	Economics

247. In 1897 the General Assembly of the State of Indiana in the USA tried to pass legislation that appears to say that is to be 3.2, though the Bill does not make it very clear. On top of that, they had the nerve to try to get everyone else to pay royalties for this discovery. Fortunately the bill didn't pass in the Senate.

248. The Fundamental Theorem of Algebra (FTA) states:—

Every polynomial equation of degree n with complex coefficients has n roots in the complex numbers.

Vieta gave equations of degree n with n roots but the first claim that there are always n solutions was made by a Flemish mathematician Albert Girard in 1629 **in Linvention en algèbre** . Gauss is usually credited with the first proof of the FTA. In his doctoral thesis of 1799, he presented his first proof and also his objections to the other proofs.

249. The German mathematician Gottfried Leibniz discovered a remarkable set of number in the from of triangle. The Leibniz numbers have a symmetry relation about vertical line. In Leibniz triangle the number in one row is obtained by adding the two numbers below it.

$$1$$

$$\frac{1}{2} \quad \frac{1}{2}$$

$$\frac{1}{3} \quad \frac{1}{6} \quad \frac{1}{3}$$

$$\frac{1}{4} \quad \frac{1}{12} \quad \frac{1}{12} \quad \frac{1}{4}$$

$$\frac{1}{5} \quad \frac{1}{20} \quad \frac{1}{30} \quad \frac{1}{20} \quad \frac{1}{5}$$

$$\frac{1}{6} \quad \frac{1}{30} \quad \frac{1}{60} \quad \frac{1}{60} \quad \frac{1}{30} \quad \frac{1}{6}$$

In the above $\frac{1}{30}+\frac{1}{20}=\frac{1}{12}$

The out side diagonal from a harmonic series.

$$1+\frac{1}{2}+\frac{1}{3}+\frac{1}{4}+\frac{1}{5}+\frac{1}{6}+......$$

250. It is interesting to know that the triangle is indispensable in building you can push a square or rectangle out of shape but not a triangle. The truss used in building is the joining together of triangles, and it seen as the components in roofs. A warren tress can take a heavy loading compared to its weight.

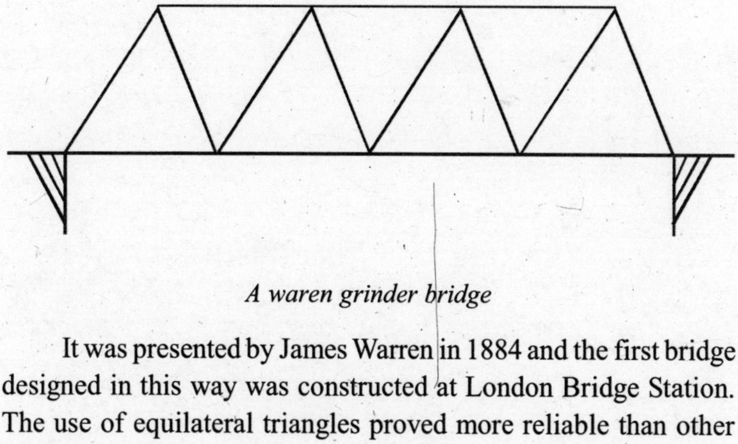

A waren grinder bridge

It was presented by James Warren in 1884 and the first bridge designed in this way was constructed at London Bridge Station. The use of equilateral triangles proved more reliable than other design.

251. The word Diagonal is a combination of two words – dia and gonus. Both the word has the Greek origin. The word dia means to pass through or join whereas the word gonus means angle. Hence

diagonal is the line segment which passes from the vertex of one angle to another in a polygon. The word diagonal was probably used in geometry by Heron of Alexandria.

□

Bibliography

1. A History of Elementary Mathematics—Florian Cajori
2. History of Mathematics—DE Smith (Volume 1 and 2)
3. Prentice Hall Encyclopedia of Mathematics—
   BeverlyHenderson West
   EllenNorma Griesbach
   JerryDuncan Taylor
   Louise ToddTaylor
4. Makers of Mathematics—Alfred Hooper
5. Mathematics Dictionary—James and James
6. Awakening India—Chinmaya Yuva Kendra
7. History of Mathematics—Lancelot Hogben
8. Outline of the History of Mathematics—Raymond Clare Archibald
9. Amusement in Mathematics—HE Dudeney
10. Fascinating Fibonacci—Trudi Hammel Garland
11. Number: Facts, Figures and Fiction—Richard Phillips
12. Mathematical Mystery Tour—Mark Wahl
13. Mathematics—David Bergamini
14. Mac Tutor Archieves
15. www.mathworld.wolfram.com

16. The Zen of Magic squares, circles and stars—C.A. Pickover
17. www.pballew.net
18. 50 Mathematical Ideas—Tony Crilly
19. Short Stories about Numbers—Rajnish Kumar
20. The Little Book of Maths, Theorem, Theories and Things—Surandera Verma

□□□